Compendium of Plant Genomes

Series Editor

Chittaranjan Kole, President, International Climate Resilient Crop Genomics Consortium (ICRCGC), President, International Phytomedomics & Nutriomics Consortium (IPNC) and President, Genome India International (GII), Kolkata, India

Whole-genome sequencing is at the cutting edge of life sciences in the new millennium. Since the first genome sequencing of the model plant *Arabidopsis thaliana* in 2000, whole genomes of about 100 plant species have been sequenced and genome sequences of several other plants are in the pipeline. Research publications on these genome initiatives are scattered on dedicated web sites and in journals with all too brief descriptions. The individual volumes elucidate the background history of the national and international genome initiatives; public and private partners involved; strategies and genomic resources and tools utilized; enumeration on the sequences and their assembly; repetitive sequences; gene annotation and genome duplication. In addition, synteny with other sequences, comparison of gene families and most importantly potential of the genome sequence information for gene pool characterization and genetic improvement of crop plants are described.

G. Craig Yencho · Bode A. Olukolu ·
Sachiko Isobe

Editors

The Sweetpotato Genome

 Springer

Editors
G. Craig Yencho (iD)
Department of Horticultural Science
North Carolina State University
Raleigh, NC, USA

Bode A. Olukolu (iD)
Department of Entomology
and Plant Pathology
University of Tennessee at Knoxville
Knoxville, TN, USA

Sachiko Isobe (iD)
Department of Frontier Research
Kazusa DNA Research Institute
Kisarazu, Chiba, Japan

ISSN 2199-4781   ISSN 2199-479X   (electronic)
Compendium of Plant Genomes
ISBN 978-3-031-65002-4   ISBN 978-3-031-65003-1   (eBook)
https://doi.org/10.1007/978-3-031-65003-1

This Springer imprint is published by the registered company Springer Nature Switzerland AG
The registered company address is: Gewerbestrasse 11, 6330 Cham, Switzerland

If disposing of this product, please recycle the paper.

*This book series is dedicated to my wife Phullara
and our children Sourav and Devleena*

Chittaranjan Kole

Preface to the Series

Genome sequencing has emerged as the leading discipline in the plant sciences coinciding with the start of the new century. For much of the twentieth century, plant geneticists were only successful in delineating putative chromosomal location, function, and changes in genes indirectly through the use of a number of "markers" physically linked to them. These included visible or morphological, cytological, protein, and molecular or DNA markers. Among them, the first DNA marker, the RFLPs, introduced a revolutionary change in plant genetics and breeding in the mid-1980s, mainly because of their infinite number and thus potential to cover maximum chromosomal regions, phenotypic neutrality, absence of epistasis, and codominant nature. An array of other hybridization-based markers, PCR-based markers, and markers based on both facilitated construction of genetic linkage maps, mapping of genes controlling simply inherited traits, and even gene clusters (QTLs) controlling polygenic traits in a large number of model and crop plants. During this period, a number of new mapping populations beyond F_2 were utilized and a number of computer programs were developed for map construction, mapping of genes, and for mapping of polygenic clusters or QTLs. Molecular markers were also used in the studies of evolution and phylogenetic relationship, genetic diversity, DNA fingerprinting, and map-based cloning. Markers tightly linked to the genes were used in crop improvement employing the so-called marker-assisted selection. These strategies of molecular genetic mapping and molecular breeding made a spectacular impact during the last one and a half decades of the twentieth century. But still they remained "indirect" approaches for elucidation and utilization of plant genomes since much of the chromosomes remained unknown and the complete chemical depiction of them was yet to be unraveled.

Physical mapping of genomes was the obvious consequence that facilitated the development of the "genomic resources" including BAC and YAC libraries to develop physical maps in some plant genomes. Subsequently, integrated genetic–physical maps were also developed in many plants. This led to the concept of structural genomics. Later on, emphasis was laid on EST and transcriptome analysis to decipher the function of the active gene sequences leading to another concept defined as functional genomics. The advent of techniques of bacteriophage gene and DNA sequencing in the 1970s was extended to facilitate sequencing of these genomic resources in the last decade of the twentieth century.

As expected, sequencing of chromosomal regions would have led to too much data to store, characterize, and utilize with the-then available computer software could handle. But the development of information technology made the life of biologists easier by leading to a swift and sweet marriage of biology and informatics, and a new subject was born—bioinformatics.

Thus, the evolution of the concepts, strategies, and tools of sequencing and bioinformatics reinforced the subject of genomics—structural and functional. Today, genome sequencing has traveled much beyond biology and involves biophysics, biochemistry, and bioinformatics!

Thanks to the efforts of both public and private agencies, genome sequencing strategies are evolving very fast, leading to cheaper, quicker, and automated techniques right from clone-by-clone and whole-genome shotgun approaches to a succession of second-generation sequencing methods. The development of software of different generations facilitated this genome sequencing. At the same time, newer concepts and strategies were emerging to handle sequencing of the complex genomes, particularly the polyploids.

It became a reality to chemically—and so directly—define plant genomes, popularly called whole-genome sequencing or simply genome sequencing.

The history of plant genome sequencing will always cite the sequencing of the genome of the model plant *Arabidopsis thaliana* in 2000 that was followed by sequencing the genome of the crop and model plant rice in 2002. Since then, the number of sequenced genomes of higher plants has been increasing exponentially, mainly due to the development of cheaper and quicker genomic techniques and, most importantly, the development of collaborative platforms such as national and international consortia involving partners from public and/or private agencies.

As I write this preface for the first volume of the new series "Compendium of Plant Genomes," a net search tells me that complete or nearly complete whole-genome sequencing of 45 crop plants, eight crop and model plants, eight model plants, 15 crop progenitors and relatives, and three basal plants is accomplished, the majority of which are in the public domain. This means that we nowadays know many of our model and crop plants chemically, i.e., directly, and we may depict them and utilize them precisely better than ever. Genome sequencing has covered all groups of crop plants. Hence, information on the precise depiction of plant genomes and the scope of their utilization are growing rapidly every day. However, the information is scattered in research articles and review papers in journals and dedicated Web pages of the consortia and databases. There is no compilation of plant genomes and the opportunity of using the information in sequence-assisted breeding or further genomic studies. This is the underlying rationale for starting this book series, with each volume dedicated to a particular plant.

Plant genome science has emerged as an important subject in academia, and the present compendium of plant genomes will be highly useful to both students and teaching faculties. Most importantly, research scientists involved in genomics research will have access to systematic deliberations on the plant genomes of their interest. Elucidation of plant genomes is of interest not only for the geneticists and breeders, but also for practitioners of an array of plant science disciplines, such as taxonomy, evolution, cytology, physiology, pathology, entomology, nematology, crop production, biochemistry, and obviously bioinformatics. It must be mentioned that information regarding each plant genome is ever-growing. The contents of the volumes of this compendium are, therefore, focusing on the basic aspects of the genomes and their utility. They include information on the academic and/or economic importance of the plants, description of their genomes from a molecular genetic and cytogenetic point of view, and the genomic resources developed. Detailed deliberations focus on the background history of the national and international genome initiatives, public and private partners involved, strategies and genomic resources and tools utilized, enumeration on the sequences and their assembly, repetitive sequences, gene annotation, and genome duplication. In addition, synteny with other sequences, comparison of gene families, and, most importantly, the potential of the genome sequence information for gene pool characterization through genotyping by sequencing (GBS) and genetic improvement of crop plants have been described. As expected, there is a lot of variation of these topics in the volumes based on the information available on the crop, model, or reference plants.

I must confess that as the series editor, it has been a daunting task for me to work on such a huge and broad knowledge base that spans so many diverse plant species. However, pioneering scientists with lifetime experience and expertise on the particular crops did excellent jobs editing the respective volumes. I myself have been a small science worker on plant genomes since the mid-1980s and that provided me the opportunity to personally know several stalwarts of plant genomics from all over the globe. Most, if not all, of the volume editors are my longtime friends and colleagues. It has been highly comfortable and enriching for me to work with them on this book series. To be honest, while working on this series I have been and will remain a student first, a science worker second, and a series editor last. And, I must express my gratitude to the volume editors and the chapter authors for providing me the opportunity to work with them on this compendium.

I also wish to mention here my thanks and gratitude to Springer staff, particularly Dr. Christina Eckey and Dr. Jutta Lindenborn, for the earlier set of volumes and presently Ing. Zuzana Bernhart for all their timely help and support.

I always had to set aside additional hours to edit books beside my professional and personal commitments—hours I could and should have given to my wife, Phullara, and our kids, Sourav and Devleena. I must mention that they not only allowed me the freedom to take away those hours from them but also offered their support in the editing job itself. I am really not sure whether my dedication of this compendium to them will suffice to do justice to their sacrifices for the interest of science and the science community.

New Delhi, India   Chittaranjan Kole

Preface

Sweetpotato, *Ipomoea batatas* (L.) Lam., is a globally important crop that has been recognized as an important source of nutrition and sustenance. It has been characterized as an orphan crop due to limited attention from global agricultural development agendas for many years, but its status is rapidly changing and there is a tremendous amount of excitement about the crop coming from multiple sectors including production, processing, value addition, marketing, and restaurants. Prior to the 2000s, investments in research, development, and promotion of sweetpotato were limited, despite its fundamental role in food security in sub-Saharan Africa and southeast Asia, and its growing importance in the higher income countries, where food options are much more abundant. The ascendence of sweetpotato into the ranks of global agricultural priorities is driven by its adaptability to diverse climatic conditions, superior nutritional value, and versatility. *The Sweetpotato Genome* highlights the growing global importance of sweet-potato, and in the following twelve chapters, leading authorities on sweet-potato improvement review how breeders, geneticists, molecular biologists, and phenomics and data management experts have worked together to advance our basic understanding of the genetics of sweetpotato. This research, which has been conducted by a comparatively small community of dedicated scientists, has enabled the creation of robust molecular marker systems and fostered the development of improved quantitative genetic theory for complex polyploids like sweetpotato that have facilitated linkage mapping and reference genome developments, which are being utilized by geneticists and breeders alike for crop improvement. These advances, and others yet to come, demonstrate that this extremely versatile crop is emerging from its status of an orphan crop, into a potentially profitable cash crop capable of providing farmers across the world with valuable food security and income-producing properties, and consumers with expanding culinary experiences.

Raleigh, USA   G. Craig Yencho
Knoxville, USA   Bode A. Olukolu
Kisarazu, Japan   Sachiko Isobe

Acknowledgments   The editors wish to acknowledge the chapter authors and the numerous administrative, laboratory, greenhouse, and field research support personnel that have worked in the background for many years to make this work possible; without their efforts, this research would not have been conducted. The authors also acknowledge the many long- and short-term investments made by organizations such as the United States Agency for International Development (USAID), the McKnight Foundation, Collaborative Crops Research Program, the Bill and Melinda Gates Foundation (BMGF) to the Consultative Group on International Agricultural Research (CGIAR) via CIP and NCSU, and the Rockefeller Foundation via the Alliance for a Green Revolution in Africa (AGRA). These grant investments have been instrumental in driving sweetpotato research forward, particularly in the developing world. Additional agencies such as the World Bank, the Food and Agriculture Organization of the United Nations (FAO), the African Development Bank (AfDB), the European Union (EU), and the United Kingdom's Foreign, Commonwealth and Development Office (FCDO) have been pivotable in enhancing the cultivation, nutritional value, and market potential of sweetpotato and they are also acknowledged. Funding through these organizations and many others not specifically named herein has highlighted the role of sweetpotato in alleviating hunger and malnutrition, particularly in the developing world. These organizations, and their many national and international partners, have played a crucial role in research and development initiatives focused on unlocking the full potential of sweetpotato as a key crop for global food security, nutrition, and sustainable agriculture.

Competing Interests   On behalf of all authors of *The Sweetpotato Genome*, the editors declare the authors have no conflicts of interest.

Contents

Contributors

Victor A. Amankwaah CSIR-Crops Research Institute, Kumasi, Ghana

Maria Andrade International Potato Center, Maputo, Mozambique

Camila Ferreira Azevedo Department of Agronomy, Federal University of Viçosa, Viçosa, Brazil

C. Robin Buell Center for Applied Genetic Technologies, Institute of Plant Breeding, Genetics and Genomics, and Department of Crop and Soil Sciences, University of Georgia, Athens, GA, USA

Qinghe Cao Xuzhou Institute of Agricultural Sciences in Jiangsu Xuhuai District, Sweetpotato Research Institute, China Agricultural Academy of Sciences, Xuzhou, China

Edward Carey International Potato Center (CIP), Kumasi, Ghana

Guilherme da Silva Pereira Department of Agronomy, Federal University of Viçosa, Viçosa, MG, Brazil

Carla Cristina da Silva Department of Agronomy, Federal University of Viçosa, Viçosa, Brazil

Gabriel de Siqueira Gesteira Department of Horticultural Science, Bioinformatics Research Center, North Carolina State University, Raleigh, NC, USA

Bryan J. Ellerbrock Clemson University, Clemson, SC, USA

Zhangjun Fei Boyce Thompson Institute, Ithaca, NY, USA;
U.S. Department of Agriculture-Agricultural Research Service, Robert W. Holley Center for Agriculture and Health, Ithaca, NY, USA

Wolfgang Grüneberg International Potato Center, Lima, Peru

John P. Hamilton Center for Applied Genetic Technologies, University of Georgia, Athens, GA, USA;
Department of Crop and Soil Sciences, University of Georgia, Athens, GA, USA

Debao Huang Department of Horticultural Science, NC State University, Raleigh, NC, USA

Sachiko Isobe Kazusa DNA Research Institute, Kisarazu, Chiba, Japan; Graduate School of Agricultural and Life Sciences, The University of Tokyo, Tokyo, Japan

Srikanth Kumar Karaikal Boyce Thompson Institute, Ithaca, NY, USA; Cornell University, Ithaca, NY, USA

Mercy Kitavi Center for Applied Genetic Technologies, University of Georgia, Athens, GA, USA;
Research Technology Support Facility (RTSF), Genomics Core, Michigan State University, East Lansing, MI, USA

Sang-Soo Kwak Plant Systems Engineering Research Center, Korea Research Institute of Bioscience and Biotechnology, Daejeon, Korea

Don LaBonte School of Plant, Environmental and Soil Sciences, 131 J.C. Miller Hall, Louisiana State University, Baton Rouge, LA, USA

Qingchang Liu Key Laboratory of Sweetpotato Biology and Biotechnology, Ministry of Agriculture and Rural Affairs, College of Agronomy and Biotechnology, China Agricultural University, Beijing, China

Wusheng Liu Department of Horticultural Science, NC State University, Raleigh, NC, USA

Chase Livengood Department of Horticultural Science, NC State University, Raleigh, NC, USA

Daifu Ma Xuzhou Institute of Agricultural Sciences in Jiangsu Xuhuai District, Sweetpotato Research Institute, China Agricultural Academy of Sciences, Xuzhou, China

Marcelo Mollinari Department of Horticultural Science, Bioinformatics Research Center, North Carolina State University, Raleigh, NC, USA

Lukas A. Mueller Boyce Thompson Institute, Ithaca, NY, USA;
Cornell University, Ithaca, NY, USA

Christine M. Nyaga Boyce Thompson Institute, Ithaca, NY, USA; Cornell University, Ithaca, NY, USA

Bonny Michael Oloka Department of Horticultural Science, North Carolina State University, Raleigh, NC, USA

Bode A. Olukolu Department of Entomology and Plant Pathology, University of Tennessee, Knoxville, TN, USA

Kenneth V. Pecota Department of Horticultural Science, NC State University, Raleigh, NC, USA

João Ricardo Bachega Feijó Rosa Federal University of Viçosa, Viçosa, Brazil

Christiano C. Simoes Boyce Thompson Institute, Ithaca, NY, USA

Olusegun Olusesan Sobowale Federal University of Viçosa, Viçosa, Brazil

Reuben Tendo Ssali International Potato Center (CIP), Kampala, Uganda

Masaru Tanaka Kyushu Okinawa Agricultural Research Center, National Agriculture and Food Research Organization, Miyakonojo, Miyazaki, Japan

Innocent Vulou Unzimai Department of Agronomy, Federal University of Viçosa, Viçosa, Brazil

Arthur Villordon LSU AgCenter Sweet Potato Research Station, Chase, LA, USA

Shan Wu Boyce Thompson Institute, Ithaca, NY, USA

Benard Yada National Crops Resources Research Institute, National Agricultural Research Organization, Kampala, Uganda

G. Craig Yencho Department of Horticultural Science, North Carolina State University, Raleigh, NC, USA

Ung-Han Yoon Genomics Division, National Institute of Agricultural Sciences, RDA, Jeonju, Korea

Zhao-Bang Zeng Department of Horticultural Science, Bioinformatics Research Center, North Carolina State University, Raleigh, NC, USA

Sweetpotato: An Orphan Crop No More?

G. Craig Yencho

Abstract

Sweetpotato, *Ipomoea batatas* (L.) Lam., a globally important crop that has been recognized as an important source of nutrition and sustenance for many years, has been characterized as an orphan crop due to limited attention from global agricultural development agendas for many years. Prior to the 2000's, investments in research, development, and promotion of sweetpotato were limited, despite its fundamental role in food security in sub-Saharan Africa and Southeast Asia. However, sweetpotato has recently ascended in the ranks of global agricultural priorities, driven by its versatility, superior nutritional value, and adaptability to diverse climatic conditions. In this chapter, the global importance of sweetpotato is reviewed with an emphasis on how breeders, geneticists, molecular biologists, and phenomics and data management experts have worked together to advance our basic understanding of the genetics sweetpotato. This research has enabled the creation of robust molecular marker systems and fostered the development of improved quantitative genetic theory for complex polyploids like sweetpotato that have facilitated linkage mapping and reference genome developments, which are being utilized by geneticists and breeders alike for crop improvement. These advances, and others yet to come, demonstrate that this extremely versatile crop is emerging from its status of an orphan crop.

Keywords

Sweet potato · Polyploids · Crop Improvement · Genomics · Phenomics

1.1 Global Importance of Sweetpotato

Sweetpotato is a critical crop in the global agricultural landscape, serving as a key source of nutrition, income, and food security for millions around the globe. Its significance is rooted in its remarkable versatility as a food, animal feed and processing crop, and its nutritional benefits, and adaptability to a wide range of environmental conditions, making it an important food staple for diverse populations, particularly in sub-Saharan Africa (SSA) and Southeast Asia. Sweetpotato has become particularly popular not only due to its health benefits but also for its culinary versatility, being used in a wide variety of dishes ranging from traditional baked

G. C. Yencho (✉)
Department of Horticultural Science,
NC State University, Raleigh, NC 27695, USA
e-mail: craig_yencho@ncsu.edu

© The Author(s) 2025
G. C. Yencho et al. (eds.), *The Sweetpotato Genome*, Compendium of Plant Genomes,
https://doi.org/10.1007/978-3-031-65003-1_1

and mashed sweetpotatoes to innovative products like sweetpotato purees, fries, chips, pies, breads, and even as an ingredient in health-conscious food products for cattle, poultry, and pet food suppliers.

The global cultivation of sweetpotato is a testimony to the crop's flexibility and its ability to integrate into various agroecological, cultural and dietary circumstances. Each geographic region produces unique varieties of sweetpotato, contributing to the crop's global genetic diversity. This diversity is not only a cultural treasure but also a critical resource for breeding programs aimed at developing new varieties that can meet future challenges.

Sweetpotato is cultivated widely across the globe, spanning tropical, subtropical, and temperate regions. It plays a pivotal role in subsistence agriculture by providing food security and nutrition, while also serving as a cash crop through the sale of fresh produce and value-added products such as snacks, flour, and animal feed in developing and developed countries. The dual role of sweetpotato enhances the economic resilience of farming communities and contributes to rural development and poverty reduction in developing and developed countries. Its ability to thrive in marginal soils and withstand harsh growing conditions, including droughts and poor soil quality, enhances sweetpotato's potential as a sustainable food source amidst the challenges of climate change and increasing global food demands. Indeed, sweetpotato is often one of the critical crops grown after devasting floods or time of drought in SSA (Loebenstein and Thottappilly 2009; Low et al. 2017; Scott 2021). The adaptability of sweetpotato to diverse environments, coupled with ongoing research and development efforts aimed at improving its yield, disease resistance, and nutritional content, ensures that this crop will continue to be a cornerstone of global food security strategies.

The global production of sweetpotato is dominated by several key countries. According to the Food and Agriculture Organization (FAO) of the United Nations, during the period 2018–2022 the top ten countries in terms of sweetpotato production were: China, Malawi, Nigeria, Tanzania, Uganda, Indonesia, Vietnam, India, United States, and Rwanda (FAO 2024; Fig. 1.1). These ten countries exemplify the global importance of sweetpotato, reflecting its role in sustaining livelihoods, enhancing nutrition, and contributing to economic stability. The International Potato Center, which has a global mandate for sweetpotato improvement and germplasm conservation as a member of the CGIAR (Consultative Group on International Agricultural Research), estimates that sweetpotatoes are grown in over 100 countries.

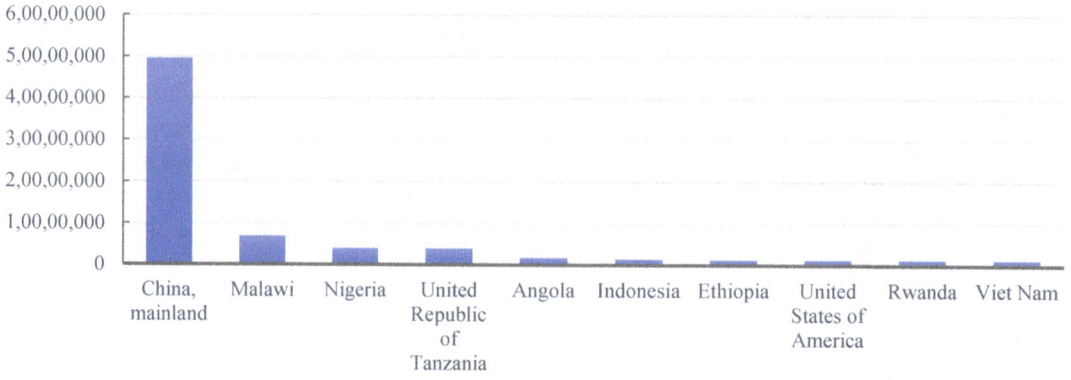

Fig. 1.1 Top ten global producers of sweetpotato (FAOSTAT 2018–2022)

China is, by far, the global leader in sweet-potato production, with the crop playing a crucial role in both the domestic and global supply of the crop (Fig. 1.2). In China, sweet-potatoes are integrated into various cuisines and snacks, and are also used for feed and industrial purposes such as the production of starch-based noodles and alcohol. In the SSA countries of Malawi, Nigeria, Tanzania, Uganda, and Ethiopia, in addition to being valued for its adaptability to different climates and soils, sweetpotato is also vital for food security and nutrition. However, it has also become an essential crop for both consumption and income, with significant production levels reflecting its rising role in the national agricultural sector in many SSA countries. In Vietnam, sweetpotato serves multiple purposes, from direct consumption to use in animal feed and starch production, making it a key agricultural commodity, while in Indonesia sweetpotato is a traditional food source that is increasingly recognized for its potential in food processing and value addition. In the United States, which predominantly produce orange-fleshed sweetpotatoes (OFSP), sweetpotatoes are considered a high value crop and they are increasingly being referred to as a super food by culinary influencers. This has increased the visibility of the crop significantly and sweetpotato production in the United States has seen significant growth over recent years, both in acreage and output, due to the increasing demand for sweetpotatoes as a nutritious and versatile food option. Currently, the U.S. is one of the largest global exporters of sweetpotato, and they are recognized for its high-quality sweetpotato cultivars, due to high quality storage and packing facilities which can provide year-round product to domestic and international markets.

1.2   Sweetpotato, a Nutritional Powerhouse and Potent Economic Driver

Sweetpotato is celebrated worldwide for its exceptional nutritional profile. Indeed, many consumer organizations consider sweetpotato to be a super food, providing a rich source of

Fig. 1.2 Production share of sweetpotato by region (FAOSTAT 2018–2022)

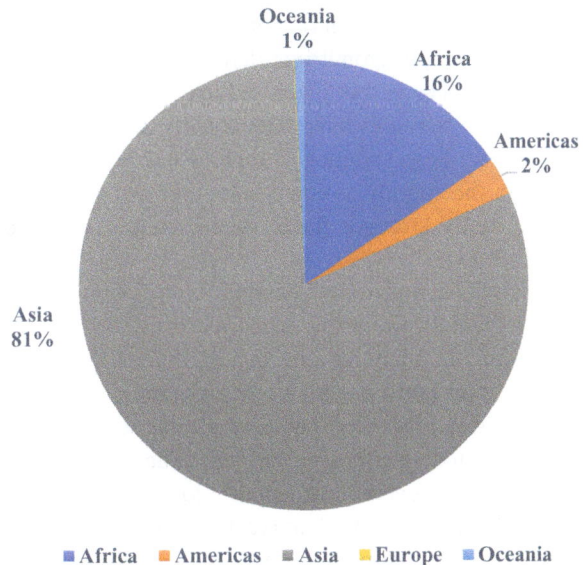

carbohydrates, dietary fibers, vitamins (especially Vitamin A in the form of beta-carotene in orange-fleshed sweetpotatoes (OFSP)), and minerals, making them an important component in the fight against malnutrition and vitamin deficiencies (Andrade et al. 2009; Scott 2021; Ojwang et al. 2023). The storage roots of sweetpotato come in a diverse array of skin and flesh colors ranging from white to cream, light yellow to deep orange, and light to dark purple. These colors provide consumers a strong visual appeal while also providing valuable health benefits due to the different phytochemicals and antioxidants present in the flesh of the sweetpotatoes, which are renowned for their nutritional benefits as they help to neutralize free radicals in the body, reducing oxidative stress and lowering the risk of chronic diseases (Truong et al. 2018). Along with this remarkable spectrum of colors, the flavor and texture profiles of sweetpotato can vary dramatically, with flavors ranging from very sweet to non-sweet and smooth, creamy to hard, mealy textures due to the starch and α- and β-amylase profiles present in the storage roots. The development and dissemination OFSP varieties have been particularly impactful in SSA, significantly improving vitamin A intake among vulnerable populations, including women of reproductive age and young children (Low et al. 2017; Girard et al. 2021). This nutritional intervention has shown remarkable efficacy in improving health outcomes and reducing morbidity from micronutrient deficiencies, however, differences in the flavor and texture profile of the OFSP varieties have undoubtedly affected adoption of these varieties in regions not accustomed to the slightly sweeter, smoother textures and generally stronger flavor characteristics of the OFSP's.

Beyond its nutritional benefits, sweetpotato plays a crucial role in the economic livelihoods of millions. In many regions, especially in SSA, SE Asia, and Latin America, sweetpotato is not only a staple food crop but also a vital source of income for smallholder farmers. The crop's short growing cycle, low input requirements, and high yield potential make it an attractive option for resource-poor farmers facing limited access to agricultural inputs and technologies. Furthermore, the burgeoning market for sweetpotato-based value-added products, ranging from processed foods such as sweetpotato fries and chips, to purees, and use as a wheat flour substitute for bread products, and industrial uses such as starch and bioethanol production, opens new avenues for economic development and diversification, contributing to rural development and poverty alleviation. In the US, the value and consumption of the crop has risen considerably during the last 20 years. Currently, the US is the largest exporter of sweetpotatoes in the world (FAOSTAT 2024), with five-year rolling averages of acres harvested and total farm-gate value of the crop increasing 70% and 224%, during the years 1996–2000 and 2019–2023, respectively (USDA-ERS 2024).

1.3 Sweetpotato Crop Resilience and Adaptability Complement Its Nutritional Status and Contribute to Sustainable Agricultural Systems

In addition to its nutritional qualities, one of the most compelling attributes of sweetpotato is its adaptability to a wide range of environmental conditions, which is well exemplified by its subtropical to temperate climate cultivation (Loebenstein and Thottappilly 2009; Grüneberg et al. 2015). An asexually propagated crop planted using un-rooted tip and stem cuttings obtained from vines and/or plant sprouts obtained from storage roots, sweetpotato exhibits remarkable resilience to various stresses, including drought and flooding, poor soil quality, and diseases. This resilience is particularly crucial for food security in vulnerable regions where the impacts of climate change threaten agricultural productivity and food availability. Its ability to grow in marginal soils with minimal water and fertilizer inputs align with the principles of sustainable agriculture, reducing the environmental footprint of food production. Additionally, sweetpotato plays a role in crop rotation and intercropping systems, contributing to soil health and biodiversity, and offering a natural pest and disease management strategy that minimizes the need for chemical inputs. These practices enhance agricultural sustainability, support the ecological balance, and preserve resources for future generations. Through breeding efforts, both traditional and genomic-assisted, new varieties of sweetpotato are being developed that further enhance this resilience, ensuring sustainable food production systems in diverse agroecological zones.

1.4 Major Milestones in Sweetpotato Improvement

Long- and short-term investments by organizations such as United States Agency for International Development (USAID), The McKnight Foundation, Collaborative Crops Research Program, the Bill and Melinda Gates Foundation (BMGF) to the CGIAR via CIP and NCSU, and the Rockefeller Foundation via AGRA (The Alliance for a Green Revolution in Africa) have been instrumental in driving sweetpotato research forward, particularly in the developing world. Additional agencies such as the World Bank, the Food and Agriculture Organization of the United Nations (FAO), the African Development Bank (AfDB), the European Union (EU), and The United Kingdom's Foreign, Commonwealth & Development Office (FCDO) have been pivotable in enhancing the cultivation, nutritional value, and market potential of sweetpotato. These investments have focused on various sweetpotato improvement needs, including improving crop yields, development and promotion of clean seed programs, efforts to enhance nutritional content, and increasing resilience to environmental stresses. Funding through these organizations and many others not specifically named herein, has highlighted the role of sweetpotato in alleviating hunger and malnutrition, particularly in the developing world. These organizations, and their many national and international partners, have played a crucial role in research and development initiatives focused on unlocking the full potential of sweetpotato as a key crop for global food security, nutrition, and sustainable agriculture. This support has enabled significant advances in breeding, pest and disease management, agronomic and seed production practices, and value chain improvements, and it has contributed to the well-being of millions of people, especially in resource-poor settings.

1.5 The Value of Traditional and Genomic-Assisted Breeding in Sweetpotato

Traditional breeding techniques have been the cornerstone of crop improvement for centuries, relying on the selection and crossing of plants to combine desirable traits. In sweetpotato, traditional breeding has typically relied on recurrent phenotypic and/or mass-selection of plants with desirable traits, such as yield, storage root size, shape and color, and resistance to local biotic and abiotic stressors. This has led to the development of varieties that are better tasting, more nutritious, and more resilient to pests and diseases (Grüneberg et al. 2015; Mwanga et al. 2017). However, in a complex functional autohexaploid like sweetpotato, traditional, phenotype-based breeding methods are very imprecise, and it often takes 8–10 years or more to develop and release a new variety. This is much too long in today's rapidly changing environments and marketplaces. Because of its polyploidy, each new segregating seedling population created through crossing results in a remarkable array of plant types and storage root shapes, colors, and qualities, which can be harnessed by sweetpotato breeding programs to create new varieties. The benefit of this in an asexually propagated crop like sweetpotato is that each genotype has the potential to be a new variety. However, sweetpotato breeding is time consuming and exceedingly difficult because it is hard to generate large populations of materials to exercise selection on as each successful cross only produces a maximum of four botanical seed, and most parents suffer from sporophytic and/or gametophytic incompatibilities, resulting in cross success rates often less than 30%. This means that it is exceedingly difficult to generate the large populations that need to be sampled to find the optimal combination of traits required for a successful new variety to be developed, and while they have resulted in advances, traditional breeding methods take too long to accomplish. Genomic-assisted breeding offers solutions to this problem.

1.6 The Rise of Genomic-Assisted Breeding in Sweetpotato

Genomic-assisted breeding has resulted in major advances in crop improvement and these tools are beginning to impact sweetpotato breeding (Mwanga, et al. 2017; Yan et al. 2022). These topics are addressed thoroughly in the following chapters of The Sweetpotato Genome. By leveraging advances in DNA sequencing, bioinformatics, and molecular biology, genomic-assisted breeding allows for the precise identification and manipulation of genes responsible for specific traits. This precision has great potential to speed the development of sweetpotato varieties with targeted improvements in a fraction of the time required by traditional methods.

Key aspects of genomic-assisted breeding include marker-assisted selection (MAS), genomic selection (GS), and gene editing techniques like CRISPR-Cas9. MAS uses molecular markers linked to desirable traits to accelerate the breeding process, while GS predicts the breeding value of individuals using genome-wide marker data. These approaches can significantly enhance the efficiency of selection for traits such as disease resistance, nutritional content, flavor and texture, and yield. Genomic-assisted breeding also facilitates the exploration of the sweetpotato's complex genome, which is particularly challenging due to its polyploid nature. Understanding the genetic basis of traits allows breeders to make more informed decisions, leading to the rapid development of superior varieties. In the future it is conceivable that

gene-editing procedures will be used to "fix" otherwise good cultivars through the introduction of improved traits via gene editing, in a fashion like backcrossing strategies used in many diploid inbred crops.

1.7 Merging Traditional and Genomic-Assisted Tools in Sweetpotato

The integration of traditional and genomic-assisted breeding tools is highly complementary with traditional breeding benefiting greatly from insights provided by genomics, such as the identification of genetic markers associated with desirable traits. An excellent example of this is the development of KASP markers for β-carotene and starch production and root knot nematode resistance in sweetpotato (Chap. 8, Fraher 2022). Conversely, genomic-assisted methods are enhanced by the empirical knowledge of plant phenotypes and environmental interactions gained through traditional breeding. Combining these approaches allows breeders to harness the full spectrum of genetic diversity within sweetpotato. This diversity is a valuable resource for introducing new traits and adapting to changing environmental conditions.

As mentioned earlier, the complex genetics of sweetpotato are a major impediment to rapid development of improved cultivars. Speed and precision can be greatly enhanced through the merging of these technologies. While traditional breeding benefits from the direct selection of phenotypes, it can be slow and is often limited by the genetic complexity of traits. Genomic-assisted breeding offers precision and speed, particularly for traits that are difficult to measure or are influenced by multiple genes, which is common in sweetpotato. The synergy of these methods can be used to accelerate the development of improved varieties (Chap. 12). Climate change, emerging pests and diseases, and the need for sustainable agricultural practices require the rapid development of adaptive, resilient, and nutritious sweetpotato varieties.

The integration of traditional and genomic-assisted breeding will enable a more dynamic response to these challenges, while leveraging the strengths of both approaches.

1.8 Future Directions and Challenges of Genomic-Assisted Breeding

The integration of traditional and genomic-assisted breeding in sweetpotato faces several challenges, including the need for capacity building in genomic technologies, especially in developing countries where sweetpotatoes are a major crop. Additionally, ethical considerations and regulatory frameworks for genetically modified organisms (GMOs) and gene-edited crops must be addressed to ensure public acceptance and market access. However, the future of sweetpotato improvement lies in the continued evolution of breeding technologies, including the potential application of artificial intelligence and machine learning to analyze complex genomic and phenotypic data. Such advancements will further enhance the precision and efficiency of sweetpotato breeding, enabling the development of varieties that are not only high-yielding and nutritious but also resilient to the challenges posed by a changing global environment.

1.9 Advances in Phenomics Complement Genomic-Assisted Breeding

Like genomics, advances in phenomics and database development are transforming the way sweetpotato breeding programs operate. Phenomics, the study of phenomes—the physical and biochemical traits of organisms as they change in response to genetic and environmental influences—provides new tools for identifying desirable traits with greater precision and speed. Phenomics has emerged as a cornerstone technology in the field of crop improvement as accurate and reliable phenotyping are so critical

to the identification of genes associated with specific traits and comprehensive understanding of how different genotypes respond to environmental stimuli.

High-throughput phenotyping technologies, such as optical imaging via various video and high-speed digital cameras systems, remote sensing using unmanned aerial vehicles (UAV's) and satellites, visible and non-visible spectroscopy methods such as near infrared spectroscopy (NIRS), high performance liquid chromatography (HPLC) and gas chromatography (GC) for chemicals enable the rapid and accurate measurement of a wide range of sweetpotato traits. These include storage root size, shape, color, nutritional content, and resistance to diseases and pests. By automating the data collection process, researchers can evaluate thousands of plants in a fraction of the time required by traditional methods, accelerating the identification and selection of superior genotypes. Through detailed phenotypic assessments, phenomics are increasingly supporting the development of sweetpotato varieties with improved nutritional profiles, such as increased levels of vitamins, minerals, and antioxidants. Simultaneously, it enables the selection of genotypes that achieve higher yields and better overall crop performance, addressing the dual challenges of nutritional security and food production efficiency.

The proliferation of genomic and phenomic data has necessitated the development of sophisticated databases to store, manage, and analyze this wealth of information. These databases have become integral to the modern crop improvement pipeline, providing a foundation for data-driven decision-making and breeding strategies (Morales et al. 2022). Advanced databases such as those described in Chap. 11 integrate phenotypic data with genomic, environmental, and agronomic information, creating a multi-dimensional resource for researchers and breeders. This integration facilitates improved understanding of the complex interactions between genotype, phenotype, and environment, enabling the identification of traits and genes associated with desirable sweetpotato characteristics. These modern database platforms are designed to be accessible and user-friendly, encouraging collaboration and data sharing among the global research community. By centralizing sweetpotato data, these platforms also foster international cooperation, facilitate the exchange of germplasm and information, and accelerate the pace of crop improvement efforts.

1.10 Summary

In summary, the future is bright for sweetpotato. The concerted efforts of the scientific community, funding organizations, and farmers, coupled with technological advancements in breeding, genomics, and food science, are paving the way for the crop's enhanced role in global agriculture. Sweetpotato has emerged as a key player in the fight against hunger and malnutrition, with its potential being increasingly recognized and leveraged across the world. The transformation of sweetpotato from an orphan crop to one of growing importance is a testament to the power of innovation, investment, and collaboration in addressing some of the most pressing challenges in food security and nutrition. However, the potential of sweetpotato is yet to be fully harnessed. Challenges such as post-harvest losses, limited access to improved varieties, and underdeveloped value chains hinder the crop's contribution to global food security and economic development. Addressing these challenges will require integrated efforts, including research and development, capacity building, and policy support. Likewise, the increasing interest in sweetpotato as a functional food, rich in health-promoting compounds, presents an opportunity for innovation in food systems (Nakitto et al. 2022). As consumer awareness about health and nutrition grows, so does the demand for foods that can deliver health benefits beyond basic nutrition. Sweetpotato, with its rich nutrient profile and potential for biofortification, stands at the forefront of this trend, offering ample opportunities for the development of novel food products and increased income opportunities for farmers globally. Clearly, sweetpotato is no longer an orphan crop, it is a crop that is ripe with economic opportunities.

Acknowledgements Funding for this research was provided by the Bill and Melinda Gates Foundation (OPP1052983 and OPP1213329) through the Genomic Tools for Sweetpotato Improvement Project (GT4SP) and Sweetpotato Genetic Advances and Innovative Seed Systems (SweetGAINS) projects, as well as the North Carolina Sweetpotato Commission, the North Carolina Crop Improvement Association (NCCIA), the North Carolina Foundation Seed Producers Inc. (NCFSP), and a Specialty Crop Research Initiative Grant No. 2021-51181-35865 from the USDA National Institute of Food and Agriculture.Conflicts of InterestThe author declares no conflict of interest.

References

Andrade M, Barker I, Cole D, Dapaah H, Elliott H, Fuentes S, Grüneberg W, Kapinga R, Kroschel J, Labarta R, Lemaga B, Loechl C, Low J, Lynam J, Mwanga R, Ortiz O, Oswald A, Thiele G (2009) Unleashing the potential of sweetpotato in Sub-Saharan Africa: current challenges and way forward. International Potato Center (CIP), Lima, Peru. Working Paper 2009-1, 197 p.

Food and Agriculture Organization (FAO) of the United Nations (2024) FAOSTAT crops and livestock products. https://www.fao.org/faostat/en/#data/QCL. Accessed 04 Apr 2024

Fraher S (2022) Advancing molecular tools for the accelerated release of root-knot nematode resistant sweetpotato varieties. MS Thesis, Dept. of Horticultural Science, North Carolina State University

Girard A, Brouwer A, Faerber E, Grant F, Low J (2021) Orange-fleshed sweetpotato: Strategies and lessons learned for achieving food security and health at scale in Sub-Saharan Africa. Open Agric 6(1):511–536. https://doi.org/10.1515/opag-2021-0034

Grüneberg WJ, Ma D, Mwanga, ROM, Carey EE, Huamani K, Diaz F, Eyzaguirre R, Guaf E, Jusuf M, Karuniawan A, Tjintokohadi K, Song Y-S, Anil SR, Hossain M, Rahaman, E, Attaluri SI, Somé K, Afuape SO, Adofo K, Lukonge E, Karanja L, Ndirigwe J, Ssemakula G, Agili S, Randrianaivoarivony JM, Chiona M, Chipungu F, Laurie SM, Ricardo J, Andrade M, Fernandes FR, Mello AS, Khan MA, Labonte DR, Yencho GC (2015) Advances in sweetpotato breeding from 1992 to 2012, pp 3–68. In: Low J, Nyongesa M, Quinn S, Parker M (eds) Potato and sweetpotato in Africa: transforming the value chains for food and nutrition security. CAB International, UK, 662 pp

Loebenstein G, Thottappilly G (eds) (2009) The sweet-potato. Springer Science+Business Media, B.V.

Low JW, Mwanga ROM, Andrade M, Carey E, Ball A-M (2017) Tackling vitamin A deficiency with biofortified sweetpotato in sub-Saharan Africa. Glob Food Sec 14:23–30. https://doi.org/10.1016/j.gfs.2017.01.004

Morales N, Ogbonna AC, Ellerbrock BJ, Bauchet GJ, Tantikanjana T, Tecle IY, Mueller LA (2022) Breedbase: a digital ecosystem for modern plant breeding. G3-Genes Gen Genet 4. https://doi.org/10.1093/g3journal/jkac078

Mwanga, ROM, Andrade MI, Carey EE, Low JW, Yencho GC, Grüneberg WJ (2017) Sweetpotato. In: Campos H, Caligari PDS (eds) Genetic improvement of tropical crops. Springer, Switzerland. 346 pp.

Nakitto M, Johanningsmeier SD, Moyo M, Bugaud C, de Kock H, Dahdouh L, Forestier-Chiron N, Ricci J, Khakasa E, Ssali RT, Mestres C (2022) Sensory guided selection criteria for breeding consumer-preferred sweetpotatoes in Uganda. Food Qual Prefer 101:104628

Ojwang SO, Okello JJ, Otieno DJ, Mutiso JM, Lindqvist-Kreuze H, Coaldrake P, Mendes T, Andrade M, Sharma N, Gruneberg W, Makunde G (2023) Targeting market segment needs with public good crop breeding investments: a case study with potato and sweetpotato focused on poverty alleviation, nutrition and gender. Front Plant Sci 14:1105079

Scott GJ (2021) A review of root, tuber and banana crops in developing countries: past, present and future. Int J Food Sci Technol 56(3):1093–1114

Truong VD, Avula R, Pecota K, Yencho, C (2018) Sweetpotato production, processing, and nutritional quality. In: Siddiq M, Uebersax MA (eds) Handbook of vegetables and vegetable processing, 2nd edn. John Wiley & Sons Ltd. https://doi.org/10.1002/9781119098935.ch35

USDA-ERS (2024) Cash receipts by state. USDA Economic Research Service. Available at: https://data.ers.usda.gov/reports.aspx?ID=17843#P75c06e7c4c1345309ddfdedba11aae3f_2_17iT0R0x33. Accessed 24 Jan 2024

Yan M, Nie H, Wang Y, Wang X, Jarret R, Zhao J, Wang H, Yang J (2022) Exploring and exploiting genetics and genomics for sweetpotato improvement: status and perspectives. Plant Commun 3(5):100332. https://doi.org/10.1016/j.xplc.2022.100332

US Efforts in Sweetpotato Genome Sequencing: Advances in the Development of Reference Genomes to Facilitate Research and Breeding of a Key Food Security Crop

2

Shan Wu, Mercy Kitavi, John P. Hamilton, C. Robin Buell, and Zhangjun Fei

Abstract

Genomic information provides a fundamental tool for modern crop breeding. Sweetpotato [*Ipomoea batatas* (L.) Lam.] is a globally important crop. However, the genome of sweetpotato is understudied due to its highly heterozygous hexaploid nature, preventing straightforward access to its genomic landscape. Here, we summarize the previous and on-going efforts in the US in the development of reference genomes for sweetpotato. Genome assemblies of diploid wild relatives, *I. trifida* and *I. triloba*, were first generated to serve as robust references for the hexaploid cultivated sweetpotato. Taking advantage of recently improved sequencing technologies and assembly algorithms, we have been generating phased genome assemblies of hexaploidy sweetpotato. Chromosome-scale haplotype-resolved genome assemblies, along with high-quality genome annotations of hexaploid sweetpotato, have been made available to the scientific and breeding communities. Multiple reference-grade phased hexaploid sweetpotato genomes set the foundation for construction of a pan-genome comprising intra- and inter-genome variations that will facilitate biological discovery and breeding of sweetpotato.

S. Wu · Z. Fei (✉)
Boyce Thompson Institute, Ithaca, NY 14853, USA
e-mail: zf25@cornell.edu

M. Kitavi · J. P. Hamilton · C. R. Buell
Center for Applied Genetic Technologies, University of Georgia, Athens, GA 30602, USA

J. P. Hamilton · C. R. Buell
Department of Crop and Soil Sciences, University of Georgia, Athens, GA 30602, USA

C. R. Buell
Institute for Plant Breeding, Genetics and Genomics, University of Georgia, Athens, GA 30602, USA

Z. Fei
U.S. Department of Agriculture-Agricultural Research Service, Robert W. Holley Center for Agriculture and Health, Ithaca, NY 14853, USA

Keywords

Sweetpotato · *Ipomoea batatas* · Hexaploid · Reference genome · Haplotype-resolved assembly

2.1 Introduction

Sweetpotato, *Ipomoea batatas* (L.) Lam. ($2n = 6x = 90$), is a globally important crop with an annual world production of more than 90 million tons over the past ten years (https://www.fao.org/

G. C. Yencho et al. (eds.), *The Sweetpotato Genome*, Compendium of Plant Genomes,
https://doi.org/10.1007/978-3-031-65003-1_2

faostat). It originated in South America (Roullier et al. 2013a; Muñoz-Rodríguez et al. 2018) and has been domesticated for more than 5000 years (Austin 1988). Sweetpotato is a rich source of carbohydrates and many other essential nutrients. Due to its hardiness and good yield, sweetpotato plays a vital role in food security to alleviate famine, especially in Africa and Southeast Asia (Loebenstein 2009). In addition, biofortification with provitamin A-rich orange-fleshed sweetpotato in sub-Saharan Africa has greatly reduced diseases caused by vitamin A deficiency in children under five years old. The World Food Prize in 2016 was awarded to scientists who pioneered this effort, highlighting the significance of sweetpotato in shifting human health outcomes.

Despite the importance of sweetpotato, its improvement has been hindered due to its genetically complex polyploid nature and the lack of robust reference genome sequences only until recently (Wu et al. 2018). Reference genome sequences of major crops and model species in early 2000s has revolutionized plant biological research and breeding (The Arabidopsis Genome Initiative 2000; International Rice Genome Sequencing Project, Matsumoto et al. 2005; Schnable et al. 2009). A surge in plant genome sequencing has enabled genomic studies using a single reference genome combined with the sequences of a large panel of individuals, which has allowed for deeper understanding of genetic diversity present in the crop species mainly at the single nucleotide polymorphism (SNP) level (Wang et al. 2018; Zhao et al. 2019). Pan-genomes that represent the genetic diversity of an entire species have greatly advanced plant breeding and evolutionary studies (Bayer et al. 2020; Della Coletta et al. 2021). More recently, having access to multiple reference-grade genome assemblies enabled by rapid advances in sequencing technologies with a goal to understand structural variants (SVs) that contribute substantially to genomic and phenotypic diversity (Alonge et al. 2020), graph-based pangenomes capturing the entire genome content including SVs of a species have been employed to facilitate the discovery of casual genetic variants of traits and the understanding of the evolution and domestication of crops (Zhou et al. 2022; Tang et al. 2022; Li et al. 2023). High-quality reference genomes and a pan-genome of hexaploid sweetpotato would provide valuable resources for genomics-enabled breeding of this important crop.

2.2 Complexity of the Hexaploid Sweetpotato Genome

Several factors have made de novo assembly of a sweetpotato genome challenging. The cultivated sweetpotato is a hexaploid with an estimated genome size of approximately 3.0 Gb (3.05 pg/2C nucleus) and with 90 chromosomes (six sets of 15 chromosomes). Polysomic inheritance observed in sweetpotato (Mollinari et al. 2020) negates an allopolyploid origin involving three divergent donors as seen in the allohexaploid (AABBDD) bread wheat, *Triticum aestivum* (International Wheat Genome Sequencing Consortium 2018). To date, the origin of the hexaploid sweetpotato remains controversial with several hypotheses proposed. An initial scenario suggested the involvement of two diploid relatives *I. trifida* ($2n = 2x = 30$) and *I. triloba* ($2n = 2x = 30$) (Austin 1988). Another hypothesis invokes autopolyploidization within *I. trifida* (Roullier et al. 2013b; Muñoz-Rodríguez et al. 2018). Morphologic (Austin 1988), cytogenetic (Srisuwan et al. 2006) and molecular evidence (Roullier et al. 2013b) have shown the close relationship between *I. trifida* and *I. batatas*. A third hypothesis suggested that *I. batatas* originated from allopolyploidization between *I. trfida* and a recently identified closely related wild autotetraploid species, *I. aequatoriensis* ($2n = 4x = 60$) (Muñoz-Rodríguez et al. 2022). Most recently, a fourth hypothesis suggested the contribution of both *I. aequatoriensis* and wild tetraploid *I. batatas* ($2n = 4x = 60$) to the hexaploid sweetpotato genome (Yan et al. 2024). These hypotheses have yet to be tested using a phased haplotype-resolved chromosome-scale hexaploid sweetpotato genome.

In addition to the high ploidy level, self-incompatibility and clonal propagation have led to high heterozygosity in the hexaploid sweetpotato genome. High heterozygosity in a genome usually leads to fragmentated genome assemblies even for diploid species, especially for short read-generated assemblies (Pryszcz and Gabaldón 2016). A previous effort attempted to utilize this highly heterozygous nature of hexaploid sweetpotato genome to generate a haplotype-resolved sweetpotato genome using short reads (Yang et al. 2017). However, due to the limitation of short reads, the resulting 836-Mb (larger than the estimated 500-Mb monoploid genome size due to redundancy) consensus genome assembly was incomplete and contained many misassemblies (Wu et al. 2018), limiting its use as the reference genome for sweetpotato.

2.3 Efforts in Sequencing the Genomes of Diploid Relatives, *I. trifida* and *I. triloba*

Smaller genome size and simpler chromosome composition of diploid relatives of polyploid crops offer substantial advantages for genomic research and breeding. Self-compatible diploid species, such as woodland strawberry (*Fragaria vesca*) (Shulaev et al. 2011) and diploid cotton *Gossypium raimondii* (Wang et al. 2012), as well as *Triticum urartu* (Ling et al. 2018) and *Aegilops tauschii* (Luo et al. 2017) (diploid progenitors of the A and D subgenomes, respectively, of the hexaploid wheat) were first sequenced to serve as reference genomes of the polyploid crops. In more complicated cases, for self-incompatible diploid progenitors, homozygous doubled monoploid lines were developed for constructing high-quality reference genomes of polyploid crops such as potato (The Potato Genome Sequencing Consortium 2011) and modern rose (Saint-Oyant et al. 2018).

While genome sequences of *I. trifida* lines were first released in 2015, they were fragmented and incomplete (Hirakawa et al. 2015). The first reference-grade genome sequences of

diploid wild relatives of cultivated hexaploid sweetpotato were reported in Wu et al. (2018), and are available at Sweetpotato Genomics Resource (http://sweetpotato.uga.edu). *I. trifida* NCNSP0306 and *I. triloba* NCNSP0323 were selected for reference genome sequencing. *I. triloba* NCNSP0323 is a highly homozygous inbred line derived from PI 618966 originally collected in Michoacan, Mexico. *I. trifida* NCNSP0306 is a self-compatible inbred line with a relatively low level of heterozygosity (0.24%) derived from PI 540724 originally collected in Magdalena, Colombia. The genomes were sequenced mainly using the Illumina short-read technology. Illumina short reads from paired-end genomic libraries and mate-pair libraries with different insert sizes ranging from 2 to 40 kb were generated and used for genome assembling, resulting in assembled scaffolds with N50 lengths of 1.2 and 6.9 Mb for *I. trifida* and *I. triloba*, respectively. PacBio long reads were also generated and used for gap-filling, and de novo-assembled BioNano maps were used to refine the assemblies. The final assemblies were 462.0 Mb and 457.8 Mb for *I. trifida* and *I. triloba*, respectively, and each was anchored and oriented onto the 15 chromosomes using a high-density genetic map. A total of 32,301 and 31,423 protein-coding genes were predicted in *I. trifida* and *I. triloba* genomes, respectively. More than 88% of the genes were assigned with putative functions by comparing their protein sequences to various public protein and domain databases.

The high-quality *I. trifida* and *I. triloba* reference genomes enabled comparative genomic analyses to uncover *Ipomoea* lineage-specific expanded gene families that function in storage root development and defense (Wu et al., 2018). Syntenic blocks within the *I. trifida* or *I. triloba* genome revealed a whole-genome triplication (WGT) event specific to the *Ipomoea* lineage that occurred around 46 million years ago (Wu et al. 2018). Functional enrichment analysis of genes induced by stress treatments suggested that the WGT event played a critical role in adaptation. Key genes associated with storage root development were also found to be contributed by this WGT event. The *I. trifida* reference was used to study the expressions of hexaploid sweetpotato orthologs that are involved in abiotic stress tolerance (Lau et al. 2018; Kitavi et al. 2023) and disease resistance (Bednarek et al. 2021), providing candidates for breeding and research. Furthermore, these diploid references facilitated the identification of genes and alleles associated with high β-carotene content in cultivated hexaploid sweetpotato (Wu et al. 2018) and the negative association between β-carotene and starch contents due to physical linkage (Gemenet et al. 2020). These findings demonstrated the robustness of the diploid *I. trifida* and *I. triloba* reference genomes in supporting studies that investigate genetic and molecular bases of sweetpotato agronomic traits.

More than 92% of genomic reads generated from the hexaploid sweetpotato could be aligned to the diploid references (Wu et al. 2018), permitting the construction of an ultra-dense phased genetic map to characterize the inheritance system in hexaploid sweetpotato (Mollinari et al. 2020). Sweetpotato accessions from the Mwanga diversity panel (MDP) have been extensively used as parents in African sweetpotato breeding programs (David et al. 2018). The MDP contains a total of 16 accessions, including breeding lines, cultivars, and landraces sourced from different areas across Uganda, as well as a few selected introduction lines. The *I. trifida* and *I. triloba* genomes were used as references to call SNPs from genome resequencing data of the 16 MDP sweetpotato lines. The resulting high-density robust polymorphic marker set confirmed the highly heterozygous nature of hexaploid sweetpotato, revealed the population structure of these key breeding lines, and improved the delineation of their phylogenetic relationships. Genomic read mapping depth of hexaploid sweetpotato to the diploid genomes also revealed chromosomal aberrations in hexaploid lines. NASPOT 5, a member of the MDP, was identified to be a double monosomic line with 88 instead of 90 chromosomes, which was confirmed using cytogenetics (Wu et al. 2018). These results showcase the usefulness of diploid reference genomes in characterizing the genetic and genomic features of hexaploid sweetpotatoes.

2.4 Development of Chromosome-Scale Haplotype-Resolved Hexaploid Sweetpotato Reference Genomes

The diploid reference genomes can serve as a fundamental tool for modern breeding of polyploid crops. However, the homozygous diploid genomes cannot fully represent the genes and allele diversity in polyploid genomes, and polyploid references are still required for detecting and studying polyploid-specific loci and genes that control agronomically important traits. Continual improvement of genome assembly algorithms and sequencing technologies that produce longer and more accurate reads combined with phase information from genetic maps have allowed for the de novo assembly of phased heterozygous diploid genomes such as those of apple and potato (Sun et al. 2020; Zhou et al. 2020). However, haplotype phasing and construction of chromosome-level assemblies of highly heterozygous autopolyploid genomes remain challenging due to the presence of more than two haplotypes with highly similar sequences. Chromosome conformation capture (Hi-C) sequencing data have been applied to resolve haplotypes and achieve chromosome-scale autopolyploid assemblies (Zhang et al. 2018, 2019; Healey et al. 2024). Yet, there is no straightforward standard method for assembling complex autopolyploid genomes. For example, in addition to PacBio high-accuracy long reads (HiFi) and Hi-C data, single-cell sequencing of diploid gametes was used to separate reads derived from different haplotypes to resolve collapsed contigs in a tetraploid potato genome assembly and reconstruct the sequences of all four haplotypes (Sun et al. 2022).

To facilitate sweetpotato breeding and research, hexaploid cultivated sweetpotato accessions, including Beauregard, Tanzania and New Kawogo, were selected for genome sequencing.

De novo assembly of these hexaploid genomes were performed using the haplotype-resolved assembler hifiasm (Cheng et al. 2021) with PacBio HiFi reads. Phased genetic maps constructed using a full-sib population derived from a cross between Tanzania and Beauregard (Mollinari et al. 2020) and another cross between Beauregard and New Kawogo (unpublished) and Hi-C data were then utilized for haplotype phasing and pseudochromosome construction. The resulting assembly of Beauregard has a total size of 2.70 Gb, with 2.54 Gb sequences (94.1% of the total assembly) anchored into haplotype-resolved 90 chromosomes. For Tanzania and New Kawogo, 2.78 Gb and 2.77 Gb were assembled with 2.53 Gb and 2.49 Gb anchored to the 90 chromosomes, respectively. Genetic map, Hi-C contact signals along the pseudochromosomes, and HiFi read alignments were further used to manually curate misassemblies. Assessment of the Beauregard, Tanzania and New Kawogo genome assemblies, including overall benchmarking universal single-copy orthologs (BUSCO) analysis (Simão et al. 2015), k-mer based evaluation of completeness, and inference of collapsed sequences in the assemblies based on read coverage, indicate that these phased assemblies are highly complete.

Genome annotation has been performed for the Beauregard, Tanzania and New Kawogo assemblies using a custom annotation pipeline for sweetpotato, resulting in the prediction of 225,111, 234,617 and 230,838 high-confidence gene models in these three assemblies, respectively, with numbers of genes in a haplome (one set of chromosomes) ranging from 34,682 to 38,505. More than 99% of the conserved plant genes were found complete in the predicted genes, indicating high completeness and quality of the genome annotation. The three genome assemblies and predicted genes have been made available to public and private research communities as a resource to facilitate sweetpotato biological discovery and breeding through Sweetpotato Genomics Resource (http://sweetpotato.uga.edu/). These hexaploid sweetpotato reference genomes provide more precise sequences for genome editing than diploid references. The phased genome assembly representing all six haplotypes in cultivated sweetpotato is crucial for studying the role of dosage and allele-specific gene expression in conferring traits of interest. In addition, through comparative genomic analyses, the chromosome-scale haplotype-resolved assemblies will be central to revealing the origin of hexaploid sweetpotato.

2.5 Towards the Development of a Sweetpotato Pan-Genome

Extensive structural variations have been found within and among tetraploid potato genomes (Hoopes et al. 2022). The same is expected for hexaploid sweetpotato genomes. Indeed, sweetpotato exhibits both intra- and inter genome structural variations. For example, our comparative analysis has detected large inversions among different haplotypes within the same sweetpotato genomes. Furthermore, aneuploidy has been discovered in cultivated sweetpotato, demonstrating an extreme form of presence/absence variation (PAV) between accessions (Wu et al. 2018).

To assist genetic analyses of complex biological traits in sweetpotato and capture causal genetic variants, a pan-genome comprising variations from different hexaploid sweetpotato accessions exhibiting contrasting traits and covering various breeding interests will be developed. A recently reported tetraploid potato pan-genome focused on the genic portion, which described the variation in gene content between haplomes as well as between accessions that result in a highly complex transcriptome in tetraploid potato (Hoopes et al. 2022). We also hypothesize that similar to potato (Hoopes et al. 2022), the clonally propagated hexaploid sweetpotato will be littered with dysfunctional and deleterious alleles, due to the inability to purge non-functional alleles via meiosis. In addition to gene PAVs, genomic variations outside genes have been found to explain a substantial proportion of phenotypic variations, and several agronomically important traits have been found to be controlled by gene regulation (Rodgers-Melnick et al. 2016; Alonge et al. 2020). Access to multiple reference-grade phased hexaploid sweetpotato genomes will provide the opportunity to construct a graph-based pan-genome to integrate intra- and inter-genomic variations from multiple and diverse accessions under a single genome coordinate system. This sweetpotato pan-genome graph will capture small and large genomic variants in both genic and intergenic regions and permit determination of their contribution to phenotypic variation. It will also serve as the foundation for biological discovery and improvement of sweetpotato breeding.

2.6   Summary

In summary, the genome sequences of diploid wild relatives *I. trifida* and *I. triloba* that we developed are useful in improving our knowledge of the mode of inheritance in hexaploid sweetpotato and the genetic basis of important agronomic traits. Recent advances in sequencing technologies and computational algorithms have allowed us to assemble the complex genome of hexaploid sweetpotatoes. The chromosome-scale haplotype-resolved hexaploid genomes present an ample opportunity for facilitating sweetpotato breeding and a deeper understanding of genetic and molecular mechanisms underlying complex traits. In the future, a pan-genome that integrates all genomic variations into a single graph will be developed. Bioinformatic tools that can effectively utilize the pan-genome to associate the variants to phenotypes will be helpful to realize the potential of these genomic resources in sweetpotato breeding.

References

Alonge M et al (2020) Major impacts of widespread structural variation on gene expression and crop improvement in tomato. Cell 182:145-161.e23

Austin DF (1988) The taxonomy, evolution and genetic diversity of sweet potatoes and related wild species. In: Exploration, maintenance and utilization of sweet potato genetic resources. Report of the first sweet potato planning conference 1987, Lima, Peru.

International Potato Centre (CIP), Lima, Peru, pp 27–59

Bayer PE, Golicz AA, Scheben A, Batley J, Edwards D (2020) Plant pan-genomes are the new reference. Nat Plants 6:914–920

Bednarek R, David M, Fuentes S, Kreuze J, Fei Z (2021) Transcriptome analysis provides insights into the responses of sweet potato to sweet potato virus disease (SPVD). Virus Res 295:198293

Cheng H, Concepcion GT, Feng X, Zhang H, Li H (2021) Haplotype-resolved de novo assembly using phased assembly graphs with hifiasm. Nat Methods 18:170–175

David MC et al (2018) Gene pool subdivision of East African sweetpotato parental material. Crop Sci 58:2302–2314

Della Coletta R, Qiu Y, Ou S, Hufford MB, Hirsch CN (2021) How the pan-genome is changing crop genomics and improvement. Genome Biol 22:3

Gemenet DC et al (2020) Quantitative trait loci and differential gene expression analyses reveal the genetic basis for negatively associated β-carotene and starch content in hexaploid sweetpotato [*Ipomoea batatas* (L.) Lam.]. Theor Appl Genet 133:23–36

Healey AL et al (2024) The complex polyploid genome architecture of sugarcane. Nature 628:804–810

Hirakawa H et al (2015) Survey of genome sequences in a wild sweet potato, *Ipomoea trifida* (H. B. K.) G. Don. DNA Res 22:171–179

Hoopes G et al (2022) Phased, chromosome-scale genome assemblies of tetraploid potato reveal a complex genome, transcriptome, and predicted proteome landscape underpinning genetic diversity. Mol Plant 15:520–536

International Wheat Genome Sequencing Consortium (2018) Shifting the limits in wheat research and breeding using a fully annotated reference genome. Science 361:eaar7191

Kitavi M et al (2023) Identification of genes associated with abiotic stress tolerance in sweetpotato using weighted gene co-expression network analysis. Plant Direct 7:e532

Lau KH et al (2018) Transcriptomic analysis of sweet potato under dehydration stress identifies candidate genes for drought tolerance. Plant Direct 2:e00092

Li N et al (2023) Super-pangenome analyses highlight genomic diversity and structural variation across wild and cultivated tomato species. Nat Genet 55:852–860

Ling HQ et al (2018) Genome sequence of the progenitor of wheat A subgenome *Triticum urartu*. Nature 557:424–428

Loebenstein G (2009) Origin, distribution and economic importance. In: The sweetpotato. Springer: Dordrecht, The Netherlands, pp 9–12

Luo MC et al (2017) Genome sequence of the progenitor of the wheat D genome *Aegilops tauschii*. Nature 551:498–502

Matsumoto T et al (2005) The map-based sequence of the rice genome. Nature 436:793–800

Mollinari M et al (2020) Unraveling the hexaploid sweetpotato inheritance using ultra-dense multilocus mapping. G3 10:281–292

Muñoz-Rodríguez P et al (2018) Reconciling conflicting phylogenies in the origin of sweet potato and dispersal to Polynesia. Curr Biol 28:1246-1256.e12

Muñoz-Rodríguez P et al (2022) Discovery and characterization of sweetpotato's closest tetraploid relative. New Phytol 234:1185–1194

Pryszcz LP, Gabaldón T (2016) Redundans: an assembly pipeline for highly heterozygous genomes. Nucleic Acids Res 44:e113

Rodgers-Melnick E, Vera DL, Bass HW, Buckler ES (2016) Open chromatin reveals the functional maize genome. Proc Natl Acad Sci USA 113:3177–3184

Roullier C, Benoit L, McKey DB, Lebot V (2013a) Historical collections reveal patterns of diffusion of sweet potato in Oceania obscured by modern plant movements and recombination. Proc Natl Acad Sci USA 110:2205–2210

Roullier C et al (2013b) Disentangling the origins of cultivated sweet potato (*Ipomoea batatas* (L.) Lam.). PLoS One 8:e62707

Saint-Oyant LH et al (2018) A high-quality genome sequence of *Rosa chinensis* to elucidate ornamental traits. Nat Plants 4:473–484

Schnable PS et al (2009) The B73 maize genome: complexity, diversity, and dynamics. Science 326:1112–1115

Shulaev V et al (2011) The genome of woodland strawberry (*Fragaria vesca*). Nat Genet 43:109–116

Simão FA, Waterhouse RM, Ioannidis P, Kriventseva EV, Zdobnov EM (2015) BUSCO: assessing genome assembly and annotation completeness with single-copy orthologs. Bioinformatics 31:3210–3212

Srisuwan S, Sihachakr D, Siljak-Yakovlev S (2006) The origin and evolution of sweet potato (Ipomoea batatas Lam.) and its wild relatives through the cytogenetic approaches. Plant Sci 171:424–433

Sun H et al (2022) Chromosome-scale and haplotype-resolved genome assembly of a tetraploid potato cultivar. Nat Genet 54:342–348

Sun X et al (2020) Phased diploid genome assemblies and pan-genomes provide insights into the genetic history of apple domestication. Nat Genet 52:1423–1432

Tang D et al (2022) Genome evolution and diversity of wild and cultivated potatoes. Nature 606:535–541

The Arabidopsis Genome Initiative (2000) Analysis of the genome sequence of the flowering plant Arabidopsis thaliana. Nature 408:796–815

The Potato Genome Sequencing Consortium (2011) Genome sequence and analysis of the tuber crop potato. Nature 475:189–195

Wang K et al (2012) The draft genome of a diploid cotton Gossypium raimondii. Nat Genet 44:1098–1103

Wang W et al (2018) Genomic variation in 3,010 diverse accessions of Asian cultivated rice. Nature 557:43–49

Wu S et al (2018) Genome sequences of two diploid wild relatives of cultivated sweetpotato reveal targets for genetic improvement. Nat Commun 9:4580

Yan M et al (2024) Haplotype-based phylogenetic analysis and population genomics uncover the origin and domestication of sweetpotato. Mol Plant 17:277–296

Yang J et al (2017) Haplotype-resolved sweet potato genome traces back its hexaploidization history. Nat Plants 3:696–703

Zhang J et al (2018) Allele-defined genome of the autopolyploid sugarcane *Saccharum spontaneum* L. Nat Genet 50:1565–1573

Zhang X, Zhang S, Zhao Q, Ming R, Tang H (2019) Assembly of allele-aware, chromosomal-scale autopolyploid genomes based on Hi-C data. Nat Plants 5:833–845

Zhao G et al (2019) A comprehensive genome variation map of melon identifies multiple domestication events and loci influencing agronomic traits. Nat Genet 51:1607–1615

Zhou Q et al (2020) Haplotype-resolved genome analyses of a heterozygous diploid potato. Nat Genet 52:1018–1023

Zhou Y et al (2022) Graph pangenome captures missing heritability and empowers tomato breeding. Nature 606:527–534

Trilateral Research Association of Sweetpotato (TRAS) *Ipomoea. trifida and I. batatas* Sequencing and Crop Improvement Efforts

Sachiko Isobe, Ung-Han Yoon, Qinghe Cao, Sang-Soo Kwak, Masaru Tanaka, Daifu Ma, and Qingchang Liu

Abstract

East Asia is an important region of sweetpotato production and consumption. To promote exchange among scientists studying sweetpotato in East Asia, the Trilateral Research Association of Sweetpotato (TRAS) was established in 2004 by sweetpotato scientists from China, South Korea, and Japan. The TRAS genome sequencing consortium was formally launched in 2014 and established a haploid-resolved and chromosome-scale de novo assembly of autohexaploid sweetpotato genome sequences. Before constructing the genome, we created chromosome-scale genome sequences in *Ipomoea trifida* using a highly homozygous accession, 'Mx23Hm', with PacBio RSII and Hi-C reads. Haploid-resolved genome assembly was performed for the sweetpotato (*I. batatas*) cultivar 'Xushu 18' by hybrid assembly with Illumina paired-end (PE) and mate-pair (MP) reads, 10X genomics reads, and PacBio RSII reads. Then, 90 chromosome-scale pseudomolecules were generated by aligning the scaffolds onto a sweetpotato linkage map. In total, 34,386 and 175,633 genes were identified on the assembled nucleic genomes of

S. Isobe (✉)
Kazusa DNA Research Institute, 2-6-7 Kazusa-Kamatari, Kisarazu 292-0818, Chiba, Japan
e-mail: sisobe@g.ccc.u-tokyo.ac.jp

U.-H. Yoon
Genomics Division, National Institute of Agricultural Sciences, RDA, 370 Nongsaengmyeong-Ro, Deokjin-Gu, Jeonju 54874, Korea

Q. Cao · D. Ma
Xuzhou Institute of Agricultural Sciences in Jiangsu Xuhuai District, Sweetpotato Research Institute, China Agricultural Academy of Sciences, Kunpeng Road 2, Jiawang District, Xuzhou 221131, China

S.-S. Kwak
Plant Systems Engineering Research Center, Korea Research Institute of Bioscience and Biotechnology, 125 Gwahak-Ro, Yuseong-Gu, Daejeon 34141, Korea

M. Tanaka
Kyushu Okinawa Agricultural Research Center, National Agriculture and Food Research Organization, Yokoichicho 6651-2, Miyakonojo 885-0091, Miyazaki, Japan

Q. Liu
Key Laboratory of Sweetpotato Biology and Biotechnology, Ministry of Agriculture and Rural Affairs, College of Agronomy and Biotechnology, China Agricultural University, Beijing 100193, China

S. Isobe
Graduate School of Agricultural and Life Sciences, The University of Tokyo, Tokyo, Japan

G. C. Yencho et al. (eds.), *The Sweetpotato Genome*, Compendium of Plant Genomes, https://doi.org/10.1007/978-3-031-65003-1_3

I. trifida and sweetpotato, respectively. The assembled genome sequences have been used for genetic and RNA-Seq analysis for agronomically important traits. The assembled genome sequences are expected to continue to contribute to genetic and genomic analysis and promote sweetpotato breeding.

Keywords

Sweetpotato · *I. trifida* · Genome · Assembly · TRAS

3.1 Introduction

Sweetpotato (*Ipomoea batatas* (L.) Lam) is widely cultivated and consumed worldwide, with a global production of 86.4 million tons in 2022 (FAO STAT). China is the leading producer, contributing 54% to the world's total production. Sweetpotato is also a popular crop in neighboring countries such as Japan and South Korea, and research on the breeding and cultivation of sweetpotato has been actively conducted in the region. To promote exchange among scientists studying sweetpotato in East Asia, the Trilateral Research Association of Sweetpotato (TRAS) was established in 2004 by sweetpotato scientists from China, South Korea, and Japan.

The inaugural symposium took place in Mokpo, South Korea, and subsequent symposiums have been held approximately every two years, rotating among the three countries. Nine international symposiums have been held to date, with the most recent one taking place in September 2022 in Xuzhou, China.

At the 5th International Sweetpotato Symposium held on Jeju Island, South Korea in 2012, agreement was reached among the three countries to undertake the construction of a reference genome for sweetpotato. After subcommittee meetings in Tokyo and Jeju in 2013, the TRAS genome sequencing consortium was formally launched in Beijing, 2014. The consortium consists of six organizations: the Jiangsu Xuzhou Sweetpotato Research Center, CAAS (China), China Agricultural University (China), Rural Development Administration (Korea), Korea Research Institute of Bioscience and Biotechnology (Korea), Institute of Sweetpotato Research, National Agriculture and Food Research Organization (Japan), and Kazusa DNA Research Institute (Japan). The composition and roles of the consortium are shown in Table 3.1.

Sweetpotato is a hexaploid species with 90 chromosomes ($2n = 6X = 90$) and a large genome size of 4.8–5.3 pg/2C nucleus (Ozias-Akins and Jarret 1994). When de novo assembly is performed in polyploid species, it is common to advance the analysis by referencing

Table 3.1 Organizational overview and role of the TRAS genome sequencing consortium

Organization	Country	Role
The Jiangsu Xuzhou Sweetpotato Research Center, CAAS	China	Development of a S1 'Xushu 18' S1 mapping population, Illumina genome sequence collection, DenovoMAGIC assembly
College of Agronomy and Biotechnology, China Agricultural University	China	'Xushu 18' preparation, Illumina genome sequence collection, DenovoMAGIC assembly
National Institute of Agricultural Sciences, RDA	Korea	PacBio and Illumina sequence collection, de novo assembly of PacBio and Illumina reads, gene prediction and annotation, organelle genome assembly
Biological Resource Center, KRIBB	Korea	Illumina genome sequence collection, DenovoMAGIC assembly
Kyushu Okinawa Agricultural Research Center, NARO	Japan	Development of a S1 'Xushu 18' S1 mapping population, *I. trifida* material preparation
Kazusa DNA Research Institute	Japan	Illumina genome sequence collection, *I. trifida* genome assembly, 'Xushu 18' S1 preparation of mapping population, DenovoMAGIC assembly, comparative analysis

the genome of closely related diploid species (Kyriakidou et al. 2018). Sweetpotato is the only species in the genus *Ipomoea* that is cultivated as a crop; among the genus's wild species, thirteen are thought to be closely related to sweetpotato (Austin 1988). Although no definitive conclusions have been reached as to the evolutionary origin and genome structure of sweetpotato, *I. trifida* (H.B.K.) Don. has been considered a likely diploid progenitor of sweetpotato (Nishiyama 1971).

In 2012, when genome sequence analysis was first proposed as an appropriate project for TRAS, the genome sequences of diploid species of *Ipomoea* had not yet been published. Therefore, the consortium decided to conduct genome analysis, not only for the hexaploid sweetpotato but also for related diploid species. For genome assembly and transcriptome analysis in *I. batatas*, we used the Chinese variety 'Xushu 18', which is a leading variety in China, bred at Xuzhou Institute of Agricultural Sciences in Jiangsu Xuhuai District and released in 1977.

3.2 Genome Assembly of *I. trifida* 'Mx23Hm'

Whole-genome sequencing and assembly was first performed for two *I. trifida* lines, a selfed line, 'Mx23Hm', and a heterozygous line, '0431–1' (Hirakawa et al. 2015). The whole-genome de novo assembly was conducted using Illumina paired-end (PE) and mate-pair (MP) libraries. The assembled genome was initially employed for genetic analysis, such as SNP detection, serving as the first reference genome for *I. trifida*. However, due to the assembly being based solely on short reads, the scaffolds exhibited fragmentation, and connectivity at the chromosomal scale was lacking.

In order to obtain chromosome-scale scaffold sequences, the RDA group obtained a total length of 64.26 Gb PacBio subreads from 'Mx23Hm' and conducted whole genome de novo assembly. De novo assembly was conducted with subreads using the SMRTMAKE assembly pipeline (Chin et al. 2013), and a total of 2881 contigs were generated with a total length of 495.7 Mb. The 2881 contigs were polished with Illumina reads, and chromosome-scale scaffolding was then performed by HiRise (Putnam et al. 2016) with 471 M Hi-C reads. The 15 chromosome-scale scaffolds and the chr0 sequences were designated as Itr_r2.2 (Table 3.2). The total length of Itr_2.2 was 502.2 Mb, including total lengths of 460.77 Mb for 15 pseudomolecules and 41.47 Mb for the chr0 scaffold. Itr_r2.2 covered 97.4% of the 'Mx23Hm' genome, when the genome size was considered to be 515.8 Mb (Hirakawa et al. 2015), while the cover ratio of the 15 chromosome-scale scaffolds was 89.3%. The ratio of complete BUSCOs was 98.5%, including 93.4% of single-copy genes and 5.1% of duplicated genes (Simão et al. 2015). The ratios of fragmented and missing BUSCOs were 0.8% and 0.7%, respectively. A total of 34,386

Table 3.2 Statistics on the assembled *I. trifida* 'Mx23Hm' (Itr_r2.2)

	Chr01–15 + Chr0	Chr01–15	Gene
Number of sequences	16	15	34,386
Total length of sequences (bp)	502,237,654	460,770,816	36,202,112
N50 length (bp)	31,779,616	31,779,616	1455
GC%	37.2	36.5	46.4
N%	1.3	0.18	0.00
BUSCO v5.2.2 embryophyta_odb10, Number of BUSCOs = 1614			
Complete	98.5	98.5	82.6
Complete single	93.4	93.5	78.5
Complete double	5.1	5	4.1
Fragmented	0.8	0.8	5.1
Missing	0.7	0.7	12.3

gene sequences were predicted on the Itr_r2.2 genome based on ab initio and evidence-based gene models.

3.3 Genome Assembly of *I. batatas* 'Xushu18'

When the TRAS genome sequencing consortium started the whole-genome de novo assembly of *I. batatas* 'Xushu 18' in 2012, long-read sequencing was expensive, and its utilization in genome assembly was not realistic. Consequently, our approach involved the use of Illumina short reads for sequencing, and the PE and MP sequences shown in Table 3.3 were obtained.

The genome size of 'Xushu 18' was estimated as 2.6 Gb on the basis of the distribution of distinct k-mers ($K = 17$) identified by jellyfish (Marçais and Kingsford 2011) with a total length of 215.7 PE read. The results of genome size estimation have varied across studies. For example, Ozias-Akins and Jarret (1994) reported that the 2C content of the sweetpotato nucleus was 4.8–5.3 pg/2C, while Srisuwan et al. (2019) reported it as 3.1–3.3 pg/2C. Given that the haploid genome size of the diploid *I. trifida* haploid

is around 500 Mb, it is reasonable to assume that the genome size of sweetpotato is around 3 Gb/2C. Therefore, it was considered that the use of jellyfish (2.6 Gb) led to an underestimation due to the influences of homologous sequences across homoeologous chromosomes.

De novo whole-genome assembly was performed with Illumina short reads using three assembly tools. However, the N50 length ranged from 347 to 1598 bp, indicating significant fragmentation (Table 3.4).

Two approaches were then used for haploid-resolved genome sequence assembly: that is, DenovoMAGIC (NRGene, Israel) for Illumina and 10X Genomics reads and Falcon-unzip (PacBio) for PacBio reads (Yoon et al. 2022). The total length of primary contigs and haplotigs was 1.8 Gb (N50 = 325.5 Kb) and 336 Mb (N50 = 44.9 Kb), respectively (Table 3.5), while total and N50 lengths assembled by DenovoMAGIC were 2.4 Gb and 2150 Kb, respectively. The shorter total lengths in PacBio and DenovoMAGIC assembly are considered to be due to the integration of sequences across homoeologous chromosomes. Consequently, hybrid assembly with the Illumina scaffolds and PacBio reads were then performed by NRGene, and a total of 110,708 sequences were generated

Table 3.3 Sequenced illumina short reads of 'Xushu 18'

Library	Platform	Insert size (base)	Total length (Gb)	Coverage (X)[a]
Paired-end	MiSeq, HiSeq	500	615.6	205
Mate-Pair	MiSeq, HiSeq	2000	23.3	8
Mate-Pair	MiSeq	5000	25.6	9
Mate-Pair	HiSeq	8000	141.9	47
Mate-Pair	MiSeq	10,000	22	7
Mate-Pair	HiSeq	15,000	135.3	45
Mate-Pair	HiSeq	20,000	145	48

[a]The coverage of the total read length obtained concerning a genome size of 3 Gb

Table 3.4 Results of de novo assembly of 'Xushu 18' with illumina short reads

Assembly tool	Newbler 3.0	SOAP de novo 2	Platanus
Number of sequences	1,443,167	8,491,291	3,112,396
Total length (bp)	985,580,805	2,662,222,474	1,213,108,214
Max length (bp)	4312	88,150	1,009,350
N50 length (bp)	1598	347	911
GC%	35.2	36.1	37.0

Table 3.5 Status of whole genome assembly in *I. batatas* 'Xushu 18'

Assemble	DenovoMAGIC + PacBio	DenovoMAGIC	Falcon unzip + Hybrid
Read types	Illumina + 10X + Pacbio	Illumina + 10X	PacBio + IIllumina
Number of sequences	110,708	163,164	6,026
Total length (bp)	2,907,440,085	2,374,736,154	1,839,235,287
N50 length (bp)	3,144,671	2,150,559	510,412

with 2.91 Gb length. The total length was close to the estimated genome size of sweetpotato, and the result suggested that hybrid assembly using Illumina DenovoMAGIC scaffolds and PacBio reads is effective for haploid-resolved assembly in autopolyploidy species.

To create chromosome-scale scaffolds, an S_1 linkage map was constructed using the variants identified on the *I. trifida* genome. The dd-RAD-Seq sequences of 437 S_1 individuals were mapped onto 520 scaffolds comprising the 'Mx23Hm' Hi-C scaffolds. A total of 534 scaffolds were aligned on the linkage map as 90 chromosome-level scaffolds. With 109,896 unplaced scaffolds, the 90 chromosome-level scaffolds were designated as IBA_r1.0. The total length of IBA_r1.0 was 2907.4 Mb, consisting of 2168.4 Mb at the chromosome level and 738.9 Mb unplaced scaffolds (Table 3.6). The ratio of complete BUSCOs assembly on IBA_r1.0 was 99.5%, including 1.7% of single-copy genes and 97.8% of duplicated genes. A total of 175,633 gene sequences were predicted for the Itr_r2.2 genome based on ab initio and evidence-based gene models.

The genome sequences of the 90 chromosome-level scaffolds were then compared with *I. trifida* genome sequences (Itr_r2.2). There was clear macro-synteny between *I. batatas* and the diploid species (Fig. 3.1).

3.4 Application of Assembled Genome Sequences for Crop Improvement and Future Prospects

The Itr_r2.2 and IBA_r1.0 genome sequences are available on Plant GARDEN (Itr_r2.2: https://plantgarden.jp/ja/list/t35884/genome/t35884.G002, IBA_r1.0: https://plantgarden.jp/ja/list/t4120/genome/t4120.G001) and have already been used for genomic and genetic analysis. For example, Suematsu et al. (2022) reported identification of a major QTL for root thickness in *I. trifida* using a QTL-Seq approach. A BC_1F_1 population derived from crosses between 'Mx23Hm' and '0431–1' was used for the analysis, and a major QTL for root thickness (*qRT1*) was identified on chr06 of the

Table 3.6 Statistics on the assembled *I. batatas* 'Xushu 18' (IBA_r1.0) genome sequences and genes

	All	Chromosome-scale	Unplaced	Gene
Number of sequences	109,986	90	109,896	175,633
Total length of sequences (bp)	2,907,375,442	2,168,449,239	738,926,203	181,148,040
N50 length (bp)	23,279,505	26,187,252	17,387	1,380.00
GC%	36.1	35.7	37.2	46.2
N%	0.07	0.14	0.34	0
BUSCO v5.2.2 embryophyta_odb10, Number of BUSCOs = 1614				
Complete	99.5	99.4	56.1	97.2
Complete single	1.7	2.5	35.5	9.5
Complete double	97.8	96.9	20.6	87.7
Fragmented	0.2	0.2	6.8	1.7
Missing	0.3	0.4	37.1	1.1

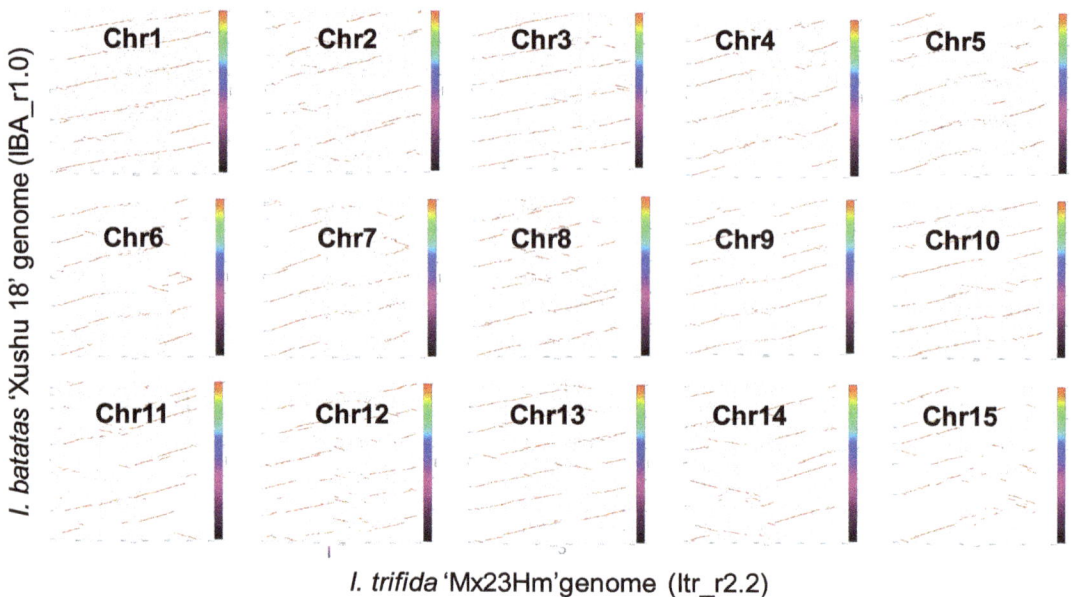

Fig. 3.1 Comparison between *I. batatas* 'Xushu 18' genome (IBA_r1.0) and *I. trifida* 'Mx23Hm' genome (Itr_r2.2) sequences

Itr_r2.2 genome. Haque et al. (2023) reported genetic analysis of starch contents (SC) using 204 F_1 progenies derived from a bi-parental cross between *I. batatas* cultivars, 'Konaishin' and 'Akemurasaki'. Base variants were identified on the Itr_r2.2 genome, and significant QTL for SC were identified on Chr15. One of the candidate genes located on the QTL regions, *IbGBSSI*, was considered to be involved in starch accumulation in sweetpotato root, by the results of qRT-PCR analysis.

For the expression analysis of starch, anthocyanin, and carotenoid genes in *I. batatas* tissues, RNA-Seq analysis was performed on RNAs extracted from the leaves at 42 days after transplantation (DAT), stems at 42 DAT, and roots at 90 DAT (Yoon et al. 2022). The fragments per kilobase of transcript per million mapped reads (FPKM) values were calculated on the genes predicted on the *I. batatas* genome, IBA_r1.0. Significantly high expressions were observed in roots for starch pathway genes. Conversely, in the leaves, the robust expression of genes associated with anthocyanin genes was observed.

Sweetpotatoes are utilized for a diverse range of purposes, including food and processed products such as starch, distilled spirits and natural colorants. Given the various applications, breeding goals for sweetpotato are diverse, necessitating genetic analyses across a multitude of traits. According to the comprehensive review by Yan et al. (2022), previous genetic analyses have predominantly focused on yield, root development, quality, and biotic resistance. Until recently, genetic analyses were predominantly conducted using the genome sequences of diploid species like *I. trifida*. However, the recent completion of the hexaploid genome sequence now paves the way for more advanced analyses. While the sweetpotato genome structure has been suggested to be either complete auto-hexaploid or auto-allo-hexaploid, elucidating the extent of genome sequence variation among homoeologous chromosomes and the conservation of gene sequences on these chromosomes is a task for the future. This advancement is anticipated to enhance our understanding of how genes governing target traits are regulated across homologous chromosomes, enabling more precise breeding strategies.

In the era of climate change, when food production faces escalating challenges, sweetpotato, with its relatively stable yields even in marginal lands, is expected to attract greater attention as a source of nutrition. The sweetpotato genomes, including those created by the TRAS consortium, are poised to serve as a crucial information resource for accelerating global sweetpotato breeding efforts. As we anticipate difficulties with food production amid changing climates, leveraging the genomic information of sweetpotato will become crucial for developing resilient crops and ensuring global food security.

Acknowledgements The TRAS genome sequencing work was performed by Dr. Ung-Han Yoon, Dr. Tae-Ho Lee, Dr. Tae-Ho Kim, Dr. Jang-Ho Hahn and Dr. Byoung Ohg Ahn of the National Institute of Agricultural Sciences, RDA, Dr. Qinghe Cao, Dr. An Zhang, Dr. Shizhuo Xiao, and Dr. Daifu Ma of the Sweetpotato Research Institute, CAAS, Dr. Hong Zhai, Dr. Xiangfeng Wang and Dr. Qingchang Liu of the College of Agronomy and Biotechnology, China Agricultural University, Dr. Ho Soo Kim, Dr. Sul-U Park, Dr. Sang-Soo Kwak of the Plant Systems Engineering Research Center, KRIBB, Dr. Masaru Tanaka, Dr. Hiroaki Tabuchi, Dr. Yoshihiro Okada, and Dr. Yasuhiro Takahata of the Kyushu Okinawa Agricultural Research Center, NARO, Dr. Jae Cheol Jeong of the Biological Resource Center, KRIBB, Dr. Soichiro Nagano of the Forest Tree Breeding Center, FFPRI, Dr. Younhee Shin of Insilicogen, Inc., Dr. Hyeong-Un Lee of the Bioenergy Crop Research Institute, National Institute of Crop Science, RDA, Dr. Seung Jae Lee of DNA Link, Inc. and College of Life Sciences and Biotechnology, Korea University, Dr. Keunpyo Lee of the Technology Cooperation Bureau, RDA, Dr. Jung-Wook Yang of the National Institute of Crop Science, RDA, and Dr. Kenta Shirasawa, Dr. Hideki Hirakawa, and Dr. Hideki Nagasaki and Dr. Sachiko Isobe of Kazusa DNA Research Institute.

References

Austin, DF (1988) The taxonomy, evolution and genetic diversity of sweet potatoes and related wild species. In: Exploration, maintenance, and utilization of sweet potato genetic resources: report of the first sweet potato planning conference 1987. International Potato Center, Lima, pp 27–59

Chin C-S, Alexander DH, Marks P et al (2013) Nonhybrid, finished microbial genome assemblies from long-read SMRT sequencing data. Nat Methods 10:563–569

FAOSTAT. https://www.fao.org/faostat/. Accessed Jan 2024

Haque E, Shirasawa K, Suematsu K et al (2023) Polyploid GWAS reveals the basis of molecular marker development for complex breeding traits including starch content in the storage roots of sweet potato. Front Plant Sci 14:1181909

Hirakawa H, Okada Y, Tabuchi H et al (2015) Survey of genome sequences in a wild sweet potato, Ipomoea trifida (H. B. K.) G. Don. DNA Res 22:171–179

Kyriakidou M, Tai HH, Anglin NL et al (2018) Current strategies of polyploid plant genome sequence assembly. Front Plant Sci 9:1660

Marçais G, Kingsford C (2011) A fast, lock-free approach for efficient parallel counting of occurrences of k-mers. Bioinformatics 27:764–770

Nishiyama I (1971) Evolution and domestication of the sweet potato. 植物学雑誌 84:377–387

Ozias-Akins P, Jarret RL (1994) Nuclear DNA content and ploidy levels in the genus ipomoea. J Am Soc Hortic Sci 119:110–115

Putnam NH, O'Connell BL, Stites JC et al (2016) Chromosome-scale shotgun assembly using an in vitro method for long-range linkage. Genome Res 26:342–350

Simão FA, Waterhouse RM, Ioannidis P et al (2015) BUSCO: assessing genome assembly and annotation completeness with single-copy orthologs. Bioinformatics 31:3210–3212

Srisuwan S, Sihachakr D, Martín J et al (2019) Change in nuclear DNA content and pollen size with polyploidisation in the sweet potato (Ipomoea batatas, Convolvulaceae) complex. Plant Biol 21:237–247

Suematsu K, Tanaka M, Isobe S (2022) Identification of a major QTL for root thickness in diploid wild sweetpotato (Ipomoea trifida) using QTL-seq. Plant Prod Sci 25:120–129

Yan M, Nie H, Wang Y et al (2022) Exploring and exploiting genetics and genomics for sweetpotato improvement: Status and perspectives. Plant Commun 3:100332

Yoon U-H, Cao Q, Shirasawa K et al (2022) Haploid-resolved and chromosome-scale genome assembly in hexa-autoploid sweetpotato (Ipomoea batatas (L.) Lam). bioRxiv 2022.12.25.521700

Evolution of Molecular Marker Use in Cultivated Sweetpotato

4

Bode A. Olukolu and G. Craig Yencho

Abstract

The use of molecular markers in sweetpotato spans first, second, and the more recent NGS-based (next-generation sequencing) third-generation platforms. This attests to the long-term interest in sweetpotato as an economically important crop. The six homoeologous chromosomes of sweetpotato lead to complex inheritance patterns that require accurate estimation of allele dosage. The use of NGS for dosage-based genotyping marked a significant advancement in sweetpotato research. Analytical pipelines have emerged to handle dosage-based genotype datasets that account for complex patterns of inheritance polyploid models. Recent approaches for dosage-based variant calling leverage reference genomes of putative ancestral progenitors or haplotype-resolved reference genome. Although pseudo-diploidized genotypes from second-generation platforms remain valuable for certain applications, especially when coarse genetic differentiation suffices, NGS-based genotyping offers a cost-effective, high-throughput, and cutting-edge alternative. Studies indicate that accurate dosage-based genotype datasets significantly enhance applications in linkage analysis, genome-wide association analysis, and genomic prediction. The affordability of NGS has spurred the adoption of high-density and dosage-sensitive molecular markers. Notably, in the three decades of molecular marker utilization in sweetpotato, about half of the peer-reviewed publications have emerged within the last four years, predominantly based on third-generation marker platforms.

Keywords

Next-generation sequencing (NGS) · Quantitative genotyping · Variant calling · Allo-autopolyploid

B. A. Olukolu (✉)
Department of Entomology and Plant Pathology, University of Tennessee, Knoxville, TN 37996-4560, USA
e-mail: bolukolu@utk.edu

G. Craig Yencho
Department of Horticultural Science, NC State University, Raleigh, NC 27695-7609, USA
e-mail: craig_yencho@ncsu.edu

4.1 Introduction

The use of molecular markers has had a significant impact on our understanding of the genetic basis for phenotypic expression and consequently enables applications in trait discovery

G. C. Yencho et al. (eds.), *The Sweetpotato Genome*, Compendium of Plant Genomes, https://doi.org/10.1007/978-3-031-65003-1_4

(i.e., identification and functional validation of marker–trait associations) and crop improvement. Landmarks in the application of molecular markers have been achieved in the past half-century and have continued to evolve (Schlotterer 2004). Applications in sweetpotato that have benefitted from the use of molecular markers include genetic diversity, DNA fingerprint, genomic prediction, genetic and physical mapping, QTL mapping, and association mapping. Like in most crops, the earliest molecular marker technologies were deployed with some limited success and resolution. Nevertheless, some of the inferences drawn from those initial efforts still hold up following the use of better technologies. In this chapter, we highlight historical perspectives on the incremental improvements of molecular markers and their applications in sweetpotato. Besides the limitations inherent in specific molecular marker platforms, we also review how the biology and genomic landscape of hexaploid sweetpotato impacts the accuracy and utility of each molecular marker system.

Sweetpotato has benefited from molecular marker technologies since the advent of the first molecular markers, i.e., protein-based Isozymes and DNA-based Restriction Fragment Length Polymorphism (RFLP) markers. Like other species where these first-generation molecular markers were used, the use of polymorphisms in DNA-based markers was rapidly favored over the protein-based Isozyme marker system. The preferences for the DNA-based markers were mostly driven by the marker density across the genome and the technical ease of developing and generating these markers. Furthermore, the relative ease of localizing DNA molecular markers to physical genomic locations enables the use of these markers for functional analysis and marker–trait validation. For example, tightly linked or functional markers can be directly used in marker-assisted selection and/or for identifying candidate genes, which is a requirement for developing genetically modified organisms. Similarly, knowledge of alleles, allele effect estimates, and allele dosage effect, particularly in polyploids, can be highly informative and useful for evaluating how much of the phenotypic variation can be explained in controlled experiments before embarking on the implementation of time-consuming selective breeding or development of transgenic lines.

The evolution of DNA molecular markers and classification into first, second, and third-generation platforms is based on a combination of marker density (i.e., low-, medium- and high-density markers, respectively) and the strategy (i.e., assay method) used for the identification of polymorphisms, i.e., DNA–DNA hybridization, PCR, and sequence-based methods, respectively. Consequently, while the properties associated with these methods are often used as the basis for classifying them into first, second, and third-generation platforms, there are exceptions where more advanced methods incorporate methods from older technologies.

First-generation molecular markers use biochemical reactions (Isozymes), hybridization of antibodies (isozymes), and DNA probes (RFLP) to detect variants of the molecule separated on a gel matrix. Second-generation platforms are based on PCR amplification with random primers (e.g., RAPD: Random Amplified Polymorphic DNA) or sequence-specific primers (e.g., SSR: Simple Sequence Repeats, SCAR: SCAR, and STS: Sequence-Tagged Sites); a combination of PCR amplification and restriction enzyme digest (e.g., AFLP: Amplified Fragment Length Polymorphism; and CAPs: Cleavage Amplified Polymorphism); and detection of allelic differences based on physical changes in the DNA conformation of amplified fragments (SSCP: Single Strand Conformation Polymorphism). Third-generation molecular marker platforms, marked by the genomic era, typically target single nucleotide polymorphisms (SNPs) by deploying assays for allele-specific hybridization (SNP chip/array and DArT: Diversity Arrays Technology) and genome-wide sequencing. The NGS-based genotyping can be untargeted (e.g., GBS: Genotyping-By-Sequencing, and ddRADseq: double digest restriction-site associated DNA sequencing) or targeted (e.g., multiplexed PCR or hybridization-based sequence capture followed by sequencing).

While third-generation platforms are designed to typically target SNPs, the start-of-the-art sequencing approaches can also identify other types of polymorphisms such as insertion-deletion polymorphisms (Indels) and short sequence repeats.

Even though sweetpotato research is supported by a small community of researchers, the diversity of molecular markers that have been used span multiple methods within each of the first-, second-, and third-generation marker platforms (Figs. 4.1 and 4.2). These highlight the

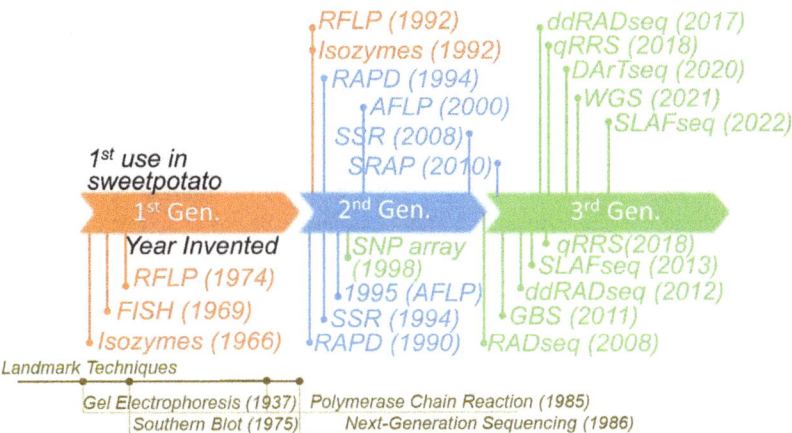

Fig. 4.1 Timeline of invention (Lewontin and Hubby 1966; John et al. 1969; Pardue and Gall 1969; Grodzicker et al 1974; Williams et al 1990; Zietkiewicz et al. 1994; Vos et al 1995; Pinkel et al. 1998; Baird et al. 2008; Elshire et al. 2011; Peterson et al. 2012; Sun et al. 2013; Wadl et al. 2018); utilization of first-; second-; and third-generation molecular markers technologies in sweetpotato and its crop wild relatives (Jarret et al. 1992; Reyes and Collins 1992; Connolly et al. 1994; Zhang et al. 2000; Veasey et al. 2008; Li et al. 2010; Shirasawa et al. 2017; Wadl et al. 2018; Bararyenya et al. 2020; Yamakawa et al. 2021; Yan et al. 2022); and landmark technologies the marker platforms depend on (Tiselius 1937; Southern 1975; Saiki et al 1985; Ronaghi et al. 1996)

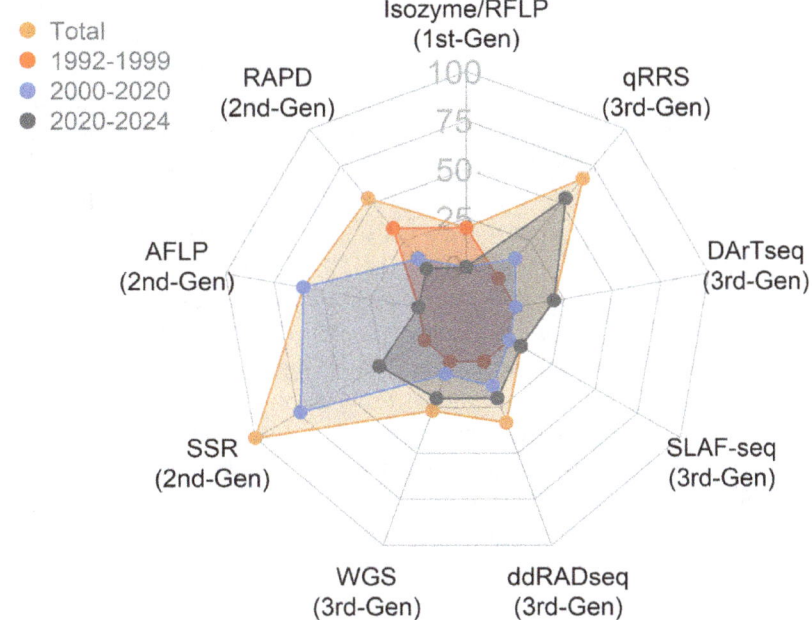

Fig. 4.2 The frequency studies published from 1992 to 2024 using various molecular marker types in *Ipomoea* spp.

importance of sweetpotato as a globally and economically important crop, as well as the interest in its polyploid evolution, domestication, and relationship with other wild diploid *Ipomoae* spp. (i.e., morning glories). These research interests span decades and pre-date molecular marker technologies.

4.2   First-Generation Molecular Marker Platforms Deployed in Sweetpotato

Following the inability to use morphological and cytogenetic markers to establish phylogenetic relationships between cultivated sweetpotato and its crop wild relatives (CWR), the earliest uses of molecular markers in sweetpotato were reported in 1992 at the USDA Agricultural Research Station, Griffin, GA, USA (Jarret et al. 1992) and North Carolina State University, Raleigh, NC, USA (Reyes and Collins 1992). The studies used RFLP and Isozyme markers, respectively, to understand phylogenetic relationships among cultivated sweetpotato and crop wild relatives, particularly species in the batatas complex. All polyploid populations were found to share almost the same number of isozymes as the diploid *I. trifida*, one of the putative ancestral progenitors of cultivated sweetpotato. The RFLP markers revealed diploid *I. trifida* and two Mexican tetraploids, an endangered *I. tabascana* (Austin 1988; Austin and De La Puente 1991; McDonald and Austin 1990) and an accession K233, that were closely related to cultivated sweetpotato with strong bootstrap support. This finding is in concordance with the fact that *I. trifida* has been observed to have traits that are required for commercial/cultivated sweetpotato, i.e., some lines develop thick roots similar to the sweetpotato, although rare, and some lines are sexually compatible with sweetpotato (Iwanaga 1988; Orjeda et al. 1990). This was reported at the first planning conference on exploration, maintenance, and utilization of sweetpotato genetic resources, held at the International Potato Center, Lima, Peru (Austin 1988). The RFLP marker data suggested two possible

rounds of polyploidization events. The study with these molecular markers provided the first evidence for the allo-autopolyploid nature of the hexaploid sweetpotato.

Besides these 2 studies that used isozyme and RFLP markers in sweetpotato (Jarret et al. 1992; Reyes and Collins 1992), another study that used chloroplast-derived RFLP markers in new world *Ipomoea* spp. (McDonald and Austin 1990). The transition to the use of second-generation markers was a rapid shift due to their ease of use and relatively lower cost. The Isozyme marker system provides a universal protocol that produces markers that are conserved across diverse genetic backgrounds (intra- and inter-specific). The evolutionary constraints on isozyme proteins and the polymorphisms that do not inactivate enzyme activity allow for their utility in intra- and inter-specific studies. The DNA-based RFLP marker and the inclusion of a more diverse set of old and new world *Ipomoea* species provided the initial insights into the origins of sweetpotato and its relationship with CWR. Nevertheless, technical issues and low-throughput assays associated with first-generation molecular markers limit their routine application. For Isozyme markers, the requirement for fresh samples (or freezing of fresh material), instability of some proteins, and a limited number of markers across the genome can limit their application. They can also suffer from bias since these proteins can be a product of direct selection that is unrelated to traits of interest or phylogenetic models of species trees (Schlotterer 2004). While RFLPs can produce a higher number of markers across the genome (i.e., a high abundance of restriction sites), in practice the tedious assay limits the number of markers that can be developed. Additionally, RFLPs and isozymes require the development of DNA probes and biochemical assays, respectively, and large amounts of starting material are required. Consequently, their applications have only been limited to diversity, phylogenetic, and low-resolution linkage mapping studies. Because they are dominant markers (2 alleles represented by the presence and absence of a band), the inability to identify heterozygotes (achievable with co-dominant markers) is a major drawback for estimating allele effects. While it is impossible to determine allele dosage with isozyme makers, allele dosage can be theoretically inferred from the intensity of bands using RFLP markers, however, this is often not feasible since there is low confidence in making such an inference.

4.3 Second-Generation Platforms Deployed in Cultivated Sweetpotato

Before the emergence of SNP arrays and NGS-based genotyping, which is central to the state-of-the-art third-generation marker platforms, second-generation molecular markers were the markers of choice since the early 1990s and remain in use. Similar to the observations in other species, only a handful of second-generation markers have been consistently used in sweetpotato (i.e., RAPD, AFLP, and SSR; Fig. 4.2). Other second-generation markers occasionally used in sweetpotato include ITS, RIP, SRAP, cpSSR, ISSR, EST-SSR, and competitive allele-specific PCR markers (e.g., KASP). While KASP markers have been sparingly used in the past, there is sustained interest to continue using this marker platform for applications that require only a few markers. Since the use and results from the low-throughput KASP marker platform have not been reported in peer-reviewed publications, its frequency of use in sweetpotato is anecdotal. Some of these second-generation markers, such as AFLPs and SSRs, can produce medium-density markers (i.e., a few thousand), which allows for their utility in genetic analyses that require genome-wide marker data. Besides their use in diversity, fingerprinting, and phylogenetic analysis, they have been used to generate the first sweetpotato linkage map and QTL analyses (Kriegner et al. 2003). The limited genome resolution in large genomes such as the approximately 3 Gb genome of sweetpotato results in major gaps in linkage groups, multiple noncontiguous linkage groups that should all map to a

single chromosome, and the unknown sequence context of the markers that are rarely anchored to positions in a physical reference genome assembly. Consequently, their utility for functional validation will tend to be limited. Second-generation markers suffer from this limitation since a primer pair can produce amplification products from multiple loci. Consequently, since amplicons are not sequenced the sequence context of alleles cannot be verified or confidently assigned to a physical genomic location. For example, SSR primer pairs tend to produce multiple alleles/bands (i.e., fragments with variable sizes) depending on PCR conditions, even in diploid genomes that should only produce 2 alleles per sample.

With the availability of reference genomes for hexaploid sweetpotato, the sequence context and physical position(s) for some of these second-generation markers can be determined. Although not limited to second-generation markers, genotypes with multiple alleles per locus and individual (even in diploids), are indicative of alleles that are obtained from multiple loci. Not to be confused with multi-allelic markers in polyploids (i.e., potentially up to 6 alleles per locus in a hexaploid), multi-locus markers derived from paralogous sequences can lead to erroneous interpretation in some genetic analyses and violate the assumption that markers are derived from a single locus. For example, this is one of the reasons for segregation distortion in genetic linkage maps and possibly false negatives and false positives during marker–trait associations. In the latter, it would be more problematic for single marker–trait genome-wide association analysis and to a lesser extent in interval mapping approaches that use interval mapping approaches based on markers that have been tested to segregate in mendelian fashion, i.e., no segregation distortion (Table 4.1).

RAPD markers were mostly used for diversity studies from 1994 to 2020, while AFLP markers were for genetic diversity and linkage analysis (genetic map construction and QTL analysis) from 2000 to 2014. Consequently, both RAPD and AFLP markers have not been

Table 4.1 Publications using second-generation molecular markers for various genetic analyses in sweetpotato

Molecular marker	Application	Performance metrics: mode of inheritance, locus specificity, level of polymorphism, and reproducibility*	References
RAPD	Genetic diversity, Phylogenetic relationship, and genetic linkage analysis	Dominant, low, medium–high, and low	Jarret et al. (1992), Connolly et al. (1994), Thompson et al. (1997), Sagredo et al. (1998), Zhang et al. (1998), Gichuki et al. (2003), Soegiant et al. (2011), Maquia et al. (2013), Yu et al. (2014), Samiyarsih et al. (2020)
AFLP	Genetic diversity, phylogenetic relationship, and genetic linkage analysis	Dominant, low, high, and high	Zhang et al. (2000), Fajardo et al. (2002), Huang et al. (2002), Kriegner et al. (2003), Zhang et al. (2004), Cervantes-Flores et al. (2008), Elameen et al. (2011), Zhao et al. (2013a), Yu et al. (2014)
ITS	Genetic diversity, and phylogenetic relationship	Dominant, medium, low, and medium	Huang et al. (2002), Huang and Sun (2000)
SSR	Genetic diversity, phylogenetic relationship, genetic linkage analysis, and parental selection	Co-dominant	Veasey et al. (2008), Chang et al. (2009), Wang et al. (2010), Yada et al. (2010), Wang et al. (2011), Roullier et al. (2013), Zhao et al. (2013a), Yu et al. (2014), Zhang et al. (2016), Kim et al. (2017), Meng et al. (2018), Lee et al. (2019), Ma et al. (2020), Meng et al. (2021), Naidoo et al. (2022), Zheng et al. (2023)
SRAP	Genetic linkage analysis		Li et al. (2010), Zhao et al. (2013b)

The performance of the marker platform is based on its use in sweetpotato and its wild relatives
Note The performance of each marker type is listed in the same order as performance metrics

reported for use in sweetpotato studies. On the other hand, SSR markers have remained in use since they were first in 2008. Application of SSR for genetic analyses includes genetic diversity analysis, phylogenetic analysis, genetic linkage analysis, marker–trait association, and genomic prediction. The first effort to characterize and develop SSR loci was based on EST-derived (expressed sequence tags) SSR markers. A study that aimed to develop genome-wide SSR markers revealed that of the 181,615 ESTs, a total of 8294 SSRs were identified from 7163 unique ESTs, i.e., a total of 3.9% of ESTs evaluated (Wang et al. 2011). The dinucleotide repeats were the predominant repeats (41.2%), with AG/CT accounting for 26.9% of repeats. Other repeats in high frequency include AAG/CTT, AT/TA, CCG/CGG, and AAT/ATT, and accounted for 13.5%, 10.6%, 5.8%, and 4.5% of SSR repeats. Consequently, only 1060 high-quality SSR primer pairs were designed. Following validation, 816 primer pairs produced reproducible and strong amplicons, while 195 and 342 SSR markers exhibited polymorphism between 2 and among 8 cultivated sweetpotato clones, respectively. The medium-density marker data derived from SSR limits their application for analysis that requires genome-wide data. The co-dominant nature and ease of use (i.e., simple PCR assay and ability to resolve some polymorphism on agarose gels) make SSR markers a popular choice.

4.3.1   Application of Second-Generation Markers for the Relationship Between Sweetpotato and Its Crop Wild Relatives, and Genetic Diversity Studies

The initial use of ITS markers derived from the internal transcribed spacer (i.e., spacer DNA situated between the subunits of ribosomal RNA genes) revealed that the ITS markers poorly resolved relationships among 13 *Ipomoea* spp (Huang and Sun 2000). In contrast, AFLP and SSR markers were found to be more efficient in characterizing genetic diversity and phylogenetic relationships at both intra- and interspecific levels in 36 accessions that represent 10 *Ipomoea* spp (Huang et al. 2002). A total of 1182 AFLP bands (loci) were identified, of which 891 were polymorphic across all accessions evaluated. The AFLP markers were generated using six primer combinations. Consistent with using first-generation markers to study the relationship between sweetpotato and its crop wild relatives, *I. trifida* was found to be the most closely related to hexaploidy sweetpotato (*I. batatas*), while *I. ramosissima* and *I. umbraticola* were the most distantly related to *I. batatas* (Huang et al. 2002). In a study that used ITS sequences, while the nuclear ITS suggested an autopolyploid origin for sweetpotato, two *I. batatas* chloroplast lineages were identified (Roullier et al. 2013). More divergence was found between the *I. batatas* chloroplast lineages than with the closest putative progenitor, *I. trifida*. While this indicated allopolyploid or all-autopolyploid origin, the study also proposed two distinct autopolyploidization events involving polymorphic wild populations of a single progenitor species. Subsequent studies with high-density third-generation molecular markers all support an allo-autopolyploid origin from previous findings.

The second-generation markers routinely used for genetic diversity studies include platforms that can produce at least medium-density markers data, i.e., hundreds to a few thousand markers (Fajardo et al. 2002; Zhang et al. 2000, 2004). A total of 210 polymorphic AFLP fragments revealed the highest genetic diversity was found in Central America and the lowest in Peru-Ecuador. These results support the hypothesis that Central America is the primary center of diversity and most likely the center of origin of sweetpotato. Furthermore, while the post-Columbus dispersal of sweetpotato to Asia and the Pacific is well documented, the hypothesis that there was a prehistoric transfer of sweetpotato by Peruvian or Polynesian voyagers from Peru to Oceania has long been a controversial issue. A set of 210 AFLP markers revealed that Mexican and Oceania cultivars grouped together, while Peru-Ecuador germplasm was genetically distant from Oceania germplasm (Zhang et al. 2004). Consequently, the study suggested that Peru-Ecuador may not be the source of the Oceania germplasm.

4.3.2 Application of Second-Generation Markers for Genetic Linkage Analysis

The medium-density marker data produced from AFLP and SSR markers have been applied to genetic linkage map construction and QTL analysis. While these molecular marker platforms cannot call dosage directly, studies have classified the pseudo-diploidized codominant SSR and dominant AFLP genotypic classes to infer dosage based on the Mendelian segregation ratio. Linkage models in these studies often tested for autopolyploidy (hexasomic) and allo-autopolyploidy (tetrasomic) without direct dosage information. Double reduction events, where sister alleles move to the same gamete during meiosis and multivalent formation, were not modeled in the era of second-generation markers. In hexasomic inheritance (autopolyploids), pairing is random with all pairs of homoeologous chromosomes. Assuming sweetpotato is an allo-autopolyploid and that preferential pairing occurs, we would expect hexasomic (if there is partial preferential pairing), tetrasomic (random pairing with pairs of 4 of 6 homologous chromosomes), and disomic (random pairing with pairs of 2 of 6 homologous chromosomes) inheritance.

Based on Jones's cytological hypotheses in sweetpotato (Jones 1967), where the other parental genotype is nulliplex, markers were classified into four types based on their segregation ratios: (1) simplex/single-dose markers present in one parent in a single copy and with a segregation ratio of 1:1 (presence: absence); (2) duplex/double-dose markers present in one parent in two copies and with hexasomic (4:1), tetrasomic (5:1), or disomic/tetradisomic (3:1) ratios; (3) triplex/triple-dose markers present in one parent in three copies and with hexasomic

(19:1), tetradisomic (11:1) or disomic (7:1) ratios; and (4) double-simplex markers present in both parents in a single copy and with a 3:1 segregation ratio (Cervantes-Flores et al. 2008; Kriegner et al. 2003). Inferring dosage or mode of inheritance in this manner is only limited to cases where the other parent is a nulliplex (or simplex in both parents), i.e., can't be applied to multiple dose marker genotypes in both parents. Furthermore, even though second-generation genetic markers can produce bridge markers (allele segregating in both parents; simplex-by-simplex marker configuration), earlier genetic linkage maps using these markers did not always use the bridge marker information to determine the 6 sets of linkage groups that correspond to the 6 sets of homoeologs. The exception is a study that partially identified some homoeologous linkage groups (Ma et al. 2020). Furthermore, no attempt was made using these markers to create a consensus map from the two parental maps. These highlight the major limitations of second-generation molecular markers for applications in genetic linkage analysis. The limitations are probably due to the inability to directly call dosage-based genotypes and the high marker density required for this genome-wide analysis.

The first attempt to construct a genetic linkage map in sweetpotato was based on a study that used 134 polymorphic markers and 76 F_1 progenies (Thompson et al. 1997). While a linkage map was not constructed, the 1:1 segregation ratio from 74 polymorphic markers (presence–absence of bands in Vardaman and Regal parents) indicated linkage map construction is possible with sufficient markers and progenies. The first genetic linkage map was constructed by (Kriegner et al. 2003) using a total of 632 (Tanzania) and 435 (Bikilamaliya) AFLP markers that were mapped to and ordered on 90 and 80 linkage groups, respectively. The map lengths covered 3655.6 cM and 3011.5 cM, respectively, with an average marker interval of 5.8 cM. In this study, to determine if sweetpotato is an autopolyploid or allopolyploid, the ratio of linkage in the coupling phase to linkage in the repulsion phase and the ratio of non-simplex to simplex markers were examined. The results support the predominance of polysomic inheritance with some degree of preferential pairing that suggests an allo-autopolyploid genome. Consequently, Cervantes-Flores et al. (2008) generated 1944 and 1751 AFLP bands in Tanzania and Beauregard, with 1511 and 1303 being single-dose markers, respectively (Cervantes-Flores et al. 2008). The framework maps consisted of 86 and 90 linkage groups for Tanzania and Beauregard, respectively. The first sweetpotato map that used SSR markers was based on ISSR (inter simple sequence repeat) markers, which produced a low-resolution map with linkage groups ranging from 10.7 to 149.1 cM (Chang et al. 2009). Only 37 and 47 markers were mapped to parental maps spanning 479.8 and 853.5 cM, respectively. Another study that used only 130 EST-SSR markers, combined it with 1824 AFLP to produce a genome-wide genetic linkage map of the sweetpotato genome (Yu et al. 2014). The only case of deploying SRAP (sequence-related amplified polymorphism) markers sweetpotato is for linkage analysis (Li et al. 2010). A total of 800 SRAP markers were used to construct 2 parental linkage maps (Luoxushu 8 × Zhengshu 20) with 473 (81 linkage groups) and 328 (66 linkage groups) spanning 5802.5 and 38967.9 cM, respectively, and with average marker interval of 10.2 and 12.0 cM, respectively.

An SSR-based genetic linkage map with high resolution that comprised only SSR was constructed by using a de novo assembly of publicly available ESTs and mRNAs in sweetpotato (Zhang et al. 2016). A total of 1824 SSR markers were obtained from 1476 primer pairs. Of these, 214 pairs of primers that identified polymorphic loci produced 1278 alleles with an average of 5.97 per locus and a major allele frequency of 0.77. Another study that mapped only 210 SSR markers produced a gene linkage map that had a low resolution and only produced small linkage groups that were limited in their representation of the sweetpotato genome (Kim et al. 2017). Similarly, a map by Ma et al. (2020) that used higher marker density (484 and

573 polymorphic SSR markers) in Jizishu 1 and Longshu 9, respectively, had a significant number of the linkage groups that mostly small and low resolution (Ma et al. 2020). Most of these efforts that use hundreds of SSR markers for linkage analysis revealed the limitation of the marker density in initial SSR markers compared to studies that show that at least a few thousand markers are required for good genome coverage.

Meng et al. (2021) were able to generate 5057 polymorphic SSR markers from 571 polymorphic genomic SSR primer pairs and 35 EST-based SSR primer pairs. They produced 90 linkage groups and covered 13,299.9 cM with a marker density of 2.6 cM. Using 3009 SSR markers, the Zhengshu 20 parental map spanned 11,1229 cM, comprised 90 linkage groups, and had a marker density of 3.7 cM. The SSR primer pairs were derived from an initial 2545 primer pairs, including 1215 genomic SSR (gSSR) primer pairs and 1330 BES-SSR (bSSR) primer pairs designed from BAC-end sequences, respectively. Using a cross between Xushu 18 and Xu 781 sweetpotato cultivars and 601 SSR primer pairs, Zheng et al. (2023) generated 5547 SSR markers and 4599 SSR markers, respectively, to produce parental maps that also spanned 18,263.5 cM and 18,043.7 cM, respectively.

At an average of 8.86–9.23 markers per SSR primer pair, it is expected that a significant number of the primer pairs are non-specifically amplified products from multiple loci (paralogous sequences). The maximum number of possible alleles per individual can be no more than 6 alleles in hexaploidy sweetpotato. To resolve this ambiguity, unique SSR bands (i.e., same fragment or similar length under low resolving power of agarose gel) are often scored as dominant markers that are derived from a single locus. Nevertheless, some paralogs can produce amplicons of the same fragment length. The ability to accurately score genotypes is crucial for polyploids since markers have a high chance to erroneously fit multiple segregation ratios under hexasomic, tetrasomic, and disomic inheritance (i.e., 1:1, 4:1, 5:1, 3:1, 19:1, 11:1, 7:1, 3:1).

4.3.3 Application of Second-Generation Markers for Selection of Parental Genotypes

While second-generation markers have not been used for genomic selection in sweetpotato, to prevent the narrowing of the genetic base during breeding, knowledge of genetic relationships and diversity is important. Naidoo et al. (2022) used SSR markers for the selection of parental cultivars to maintain genetic diversity during breeding (Naidoo et al. 2022). Using 31 genotypes originating from the African and American continent and eight highly polymorphic SSR primers that produced 83 alleles, it was revealed that despite the high diversity among the genotypes, genetic distances among the genotypes were relatively low. To some extent, clustering identified three groups that reflect geographic origins and pedigree. The study suggested two heterotic groups African and American origin.

4.4 Third-Generation Platforms Deployed in Cultivated Sweetpotato

The third-generation molecular markers (aka next-generation or advanced molecular markers) are the current state-of-the-art that represents a significant advancement marked by high-throughput genotyping at a significantly lower cost than any of the other marker platforms. These molecular marker platforms use technologies that include DNA arrays (on slides or beads made from plastic or glass) or NGS. They offer improved capabilities for applications that require high-density genome-wide marker data, and allele-dose information, and for studying complex traits and complex genomic features.

Diploid and a few polyploid organisms started benefiting from early third-generation marker platforms, including SNP arrays developed in 1998 (The Whitehead Institute and Affymetrix SNP array/chip; Pinkel et al. 1998) and NGS-based genotyping (Baird et al. 2008; Balagué-Dobón et al. 2022; Elshire et al. 2011;

Peterson et al. 2012). However, sweetpotato, like most complex polyploids lagged due to the prohibitive cost of developing SNP arrays for a small community of use and the inability of early NGS-based genotyping platforms to accurately capture allele dosage. Initially, the ability of SNP arrays to call allele dosage was limited and in the cases where they were used in polyploids, genotype calls were often limited to $2 \times$ pseudo-diploidized genotypes in autopolyploids (e.g., potato) or subgenome-specific diploid genotype calls in allopolyploid (e.g., wheat) (Sun et al. 2020). Since then, the development of tools such as FitTetra (Zych et al. 2019) and ClusterCall (Schmitz Carley et al. 2017) now allow for dosage calling, although the application is limited to auto-tetraploids.

For higher ploidy levels in autopolyploids, the development of superMASSA allowed for dosage calling using a graphical Bayesian model for SNP genotyping. However, its application is limited to biparental populations since it requires Mendelian segregation information to re-classify genotypes into appropriate dosage classes (Serang et al. 2012). Although it can model all dosage configurations, it is similar to earlier approaches used for second-generation markers in that dosage calls are imputed or re-assigned based on expected Mendelian segregation rather than strictly using allelic read depth information. The approach is necessary since early third-generation marker platforms are inherently limited in their ability to accurately quantify allele dose. The polyrad (Clark et al. 2019) and updog (Gerard et al. 2018) software provide similar functionality as superMASSA but extend dosage calling to diversity panels and natural populations by using multiple features to update priors. The features in polyrad include population structure, model allele frequency gradients, rate of self-fertilization range from zero to one, and linkage disequilibrium of markers that have known physical map positions in the reference genome. The features in updog include allele bias, overdispersion, and sequencing error.

While the advent of reduced representation sequencing (RRS) democratized high throughput genotyping in model, non-model, and under-studied crops, the ability to use them for dosage calling is limited in the first iteration of RRS protocols (e.g., RADseq, GBS, and ddRADseq). This is due to the allele read depth ratios that are often highly skewed and lack uniformity in read depth across the genome resulting in a significant number of loci with low read depth that is insufficient for dosage calling. These limitations necessitated the need for tools such as superMASSA, polyrad, and updog. In studies where the first iteration of RRS methods was used, sweetpotato genotype calls were often based on $2 \times$ pseudo-diploidized genotype calls rather than $6 \times$ dosage calls. Examples of sweetpotato studies that used pseudo-diploid genotypes from RRS data include using DArTseq for GWAS, quality assurance and control, and genomic prediction studies (Bararyenya et al. 2020; Gemenet et al. 2020a, 2020b); and using SLAF-seq for linkage/QTL analysis (Yan et al. 2022). While analysis with the pseudo-diploidized genotypes produced meaningful results, comparison with $6 \times$ dosage genotype calls in these studies revealed using dosage calls often produced superior results (Gemenet et al. 2020b). Other studies have attempted to use this first iteration RRS method with some limited success (Table 4.2).

Dosage-sensitive genotyping-by-sequencing (qRRS-based genotyping) has emerged as an amenable and robust strategy in polyploids due to its low cost and ability to quantitatively sequence alleles for dosage estimation. While approaches implementing quantitative reduced representation sequencing, GBSpoly and OmeSeq-qRRS, have been used in sweetpotato (da Silva et al. 2020; Gemenet et al. 2020b; Mollinari et al. 2020; Oloka et al. 2021; Wadl et al. 2018; Wu et al. 2018), multiplexed-PCR based approaches are been tested in sweetpotato and other polyploid crops. The GBSpoly protocol is a modification of the ligation-based GBS protocol that uses double-digestion with methylation-sensitive (*TseI*; rare cutter) and methylation-insensitive (*CviAII*, frequent cutter) restriction enzymes. To improve the quantitative sequencing assay, a library construction method (OmeSeq-qRRS) uses isothermal amplification (instead of a ligation approach) to incorporate barcoded adapters into genomic fragments following double-digestion of the genome with methylation-insensitive restriction enzymes (*NsiI* and *NlaII*). The methylation-insensitive restriction enzymes eliminate variability in hypomethylated and hypermethylated sequences, which are also variable across tissue types. A lower number of overall NGS reads is sufficient to achieve similar marker density in OmeSeq-qRRS compared to GBSpoly-qRRS. Additionally, the ligation bias of smaller

Table 4.2 Publications using third-generation molecular markers for various genetic analyses in sweetpotato

Molecular marker	Application	References
ddRADseq	Genetic linkage analysis, GWAS	Shirasawa et al. (2017), Okada et al. (2019), Haque et al. (2020a), Haque et al. (2020b), Yamakawa et al. (2021), Obata et al. (2022)
DArTseq	GWAS, genomic prediction, and quality assurance and control	Bararyenya et al. (2020), Gemenet et al. (2020a, b)
WGS	Genetic diversity, QTL-seq, GWAS, and phylogenetic relationship	Yamakawa et al. (2021), Munoz-Rodriguez et al. (2022), Wu et al. (2018), Xiao et al. (2023), Yan et al. (2024)
qRRS	Genetic diversity, genetic linkage analysis, and genomic prediction	Wadl et al. (2018), Wu et al. (2018), da Silva et al. (2020), Gemenet et al. (2020b), Mollinari et al. (2020), Oloka et al. (2021), Batista et al. (2022), Slonecki et al. (2023)
SLAF-seq	Genetic linkage analysis	Yan et al. (2022)

The performance of the marker platform is shown based on its use in sweetpotato and its wild relatives
The performance of each marker type is listed in the same order as performance metrics

genomic fragments and chimeric ligation are mitigated and eliminated, respectively, by using isothermal amplification.

The RRS/NGS-based genotyping is more amenable to polyploids than SNP arrays, which require significant assay development costs and accurate genome sequence data and assembly. The more recent NGS-based genotyping platforms aim to quantitatively sequence loci while maintaining allelic ratio in other to estimate dosage more accurately. Consequently, tools such as GATK, Freebayes, and Freebayes directly use quantitative sequencing information (i.e., allele read depth) for dosage-based variant calling by using haplotype-based calling. Variant calling based on dosage presents greater challenges in polyploids due to many potential genotypes at each locus, which arise from various combinations of unique alleles. Sequence reads are incapable of distinguishing identical copies of alleles, particularly when not physically linked to other heterogenous alleles. Additionally, as the ploidy level increases, determining allele copy numbers through dosage calling becomes increasingly complex at a specific read depth. The adoption of a haplotype-based strategy enhances genotyping accuracy by jointly assessing the combinations of multiple nearby alleles, known as haplotypes (Cooke et al. 2021).

While variant calling tools for polyploids assume that the genome is autopolyploid, studies in sweetpotato often indicate that it is an allo-autopolyploid and that large structural variations might exist between sweetpotato accessions (Wu et al. 2018). To address the allo-autopolyploid nature, accurate variant calling in polyploids can benefit from using multiple reference genomes of putative ancestral progenitors and haplotype-resolved genome assembly (6 sweetpotato haplomes). In addition to the sequencing of putative ancestral progenitors within the *I. batatas* complex (*I. trifida*, *I. triloba*, and *I. tabascana*), haplotype-resolved genome assemblies based on multiple sweetpotato cultivars are available (http://sweetpotato.uga.edu). A variant calling pipeline, GBSapp (Bararyenya et al. 2020; Gemenet et al. 2020b; Wadl et al. 2018), was developed to address the allo-autopolyploid nature of the sweetpotato genome by

resolving sequence reads that map uniquely to haplomes or subgenomes and that are conserved across all 6 sweetpotato haplomes or subgenomes (using putative ancestral progenitors). Modeling dosage based on the number of homoeologs containing a specific sequence is particularly important for allo-autopolyploids since current variant calling tools, including haplotype-based variant calling tools, will assume and erroneously coarse genotypes to the dosage specified by the user (i.e., assumes strict autopolyploidy). In the absence of high-quality and complete haplotype-resolved reference genome assembly, which is preferred for variant calling, the GBSapp pipeline can use the known progenitors. Since the identity of the other ancestral progenitor is not known with certainty, GBSapp uses the closest ancestral diploid progenitor (*I. trifida*) and the more distantly related diploid (*I. triloba*) as reference subgenomes. This ensures that despite the evolutionary divergence of the latter, sequences conserved across both genomes would likely exist hexaploid sweetpotato and other species within the batatas species complex, hence dosage would be most likely 6 × dose.

4.4.1 Application of Third-Generation Markers for the Relationship Between Sweetpotato and Its Crop Wild Relatives, and Genetic Diversity Studies

The two studies conducted to understand the origins and domestication of sweetpotato were based on shotgun whole genome sequencing (Munoz-Rodriguez et al. 2022; Yan et al. 2024). Both studies confirmed *I. trifida* as a putative ancestral progenitor, that hexaploid sweetpotato is an allo-autopolyploid, and aimed to identify the putative tetraploid ancestral progenitor. These are findings that significantly advance the understanding of sweetpotato domestication and that inform if variant calling, and other genetic analyses should model sweetpotato as an autopolyploid or allo-autopolyploid. Munoz-Rodriguez et al. (2022) proposed *I. aequatoriensis*, a species of Mexican origin, played a direct role in the origin of the hexaploidy sweetpotato. This also underscored

Central America as the origin of sweetpotato. Likewise, Yan et al. (2024) also used a haplotype-based phylogenetic analysis (HPA) to confirm that sweetpotato originated from reciprocal crosses between a diploid and tetraploid progenitor.

Genetic diversity analysis performed with two sets of USDA diversity panel (417 and 604 accessions) and using the GBSpoly-qRRS marker platform revealed similar results (Slonecki et al. 2023; Wadl et al. 2018). The clusters identified from STRUCTURE and phylogenetic analysis correspond to the geographical location that the accessions were collected from. Accessions from the Pacific Islands and Caribbean/Central American cluster within close proximity, supporting initial studies that germplasm from the Pacific Islands originated from Central America. Distinct clusters comprised of North American accessions cluster within close proximity.

4.4.2 Application of Second-Generation Markers for Genetic Linkage Analysis

Genetic linkage maps constructed with third-generation markers are typically marked by of high-density marker data set (i.e., about 30,000 markers). Early NGS-based genotyping platforms (ddRADseq and SLAF-seq) are limited to using simplex and nulliplex markers for linkage analysis (Shirasawa et al. 2017), while recent quantitative NGS-based genotyping markers (qRRS) using all marker configurations include low- and high-dose markers (Mollinari et al. 2020). Several publications have used third-generation markers for both QTL and genome-wide analysis (Bararyenya et al. 2020; da Silva et al. 2020; Gemenet et al. 2020c; Haque et al. 2020a; Haque et al. 2020b; Oloka et al. 2021).

4.4.3 Application of Second-Generation Markers in Genomic Prediction

The third-generation markers deployed for genomic prediction in sweetpotato are derived from the DArTSeq and GBSpoly-qRRS platforms (Batista et al. 2022; Gemenet et al. 2020b). Using DArtSeq (pseudo-diploidized markers) and GBSpoly-qRRS, genomic predictive abilities (PA) in a biparental population (Beauregard x Tanzania, BT) across root quality and yield-related traits revealed that models that used allele dosage information and G-matrix based additive effects have the best PA for most of the traits (Gemenet et al. 2020b).

4.5 Conclusion

Until 2018, sweetpotato lagged in the use of high throughput third-generation molecular markers since existing methods predating 2018 were limited in their ability to estimate allele dosage. Consequently, the need for high throughput and inexpensive genotyping in sweetpotato has driven innovations in quantitative genotyping-by-sequencing (or qRRS). The availability of NGS-based RRS and shotgun whole genome sequencing (WGS) data in sweetpotato has prompted the need to develop cutting-edge bioinformatic and analytical tools/pipelines for polyploid genetics that are being applied to other simple and complex polyploids. These tools include GBSapp (Wadl et al. 2018), MAPpoly (Mollinari et al. 2020), QTLpoly (da Silva et al. 2020), VIEWpoly (Taniguti et al. 2022), haplotype-based phylogenetic analysis (Yan et al. 2024), Ranbow (Moeinzadeh et al. 2020), and ngsAssocPoly (Yamamoto et al. 2020). The emergence of molecular marker platforms sensitive enough for quantifying allele dosage is revitalizing various areas of research in sweet potatoes and their wild counterparts.

Acknowledgements Funding for this research was provided by the Bill and Melinda Gates Foundation (OPP1052983 and OPP1213329) through the Genomic Tools for Sweetpotato Improvement Project (GT4SP) and Sweetpotato Genetic Advances and Innovative Seed Systems (SweetGAINS) projects, as well as a Specialty Crop Research Initiative Grant No. 2021-51181-35865 from the USDA National Institute of Food and Agriculture.

Conflicts of Interest The authors declare no conflict of interest.

References

Austin DF (1988) The taxonomy, evolution and genetic diversity of sweet potatoes and related wild species. In: Gregory P (ed) Exploration, maintenance and utilization of sweet potato genetic resources. Report of the 1st sweet potato planning conference 1987, International Potato Center, Lima, Peru

Austin DF, De La Puente F (1991) *Ipomoea tabascana*, an endangered tropical species. Econ Bot 45:435. https://doi.org/10.1007/BF02887086

Baird NA, Etter PD, Atwood TS, Currey MC, Shiver AL, Lewis ZA, Selker EU, Cresko WA, Johnson EA (2008) Rapid SNP discovery and genetic mapping using sequenced RAD markers. PLoS ONE 3(10):e3376. https://doi.org/10.1371/journal.pone.0003376

Balagué-Dobón L, Cáceres A, González JR (2022) Fully exploiting SNP arrays: a systematic review on the tools to extract underlying genomic structure. Brief Bioinform 23(2). https://doi.org/10.1093/bib/bbac043

Bararyenya A, Olukolu BA, Tukamuhabwa P, Grüneberg WJ, Ekaya W, Low J, Ochwo-Ssemakula M, Odong TL, Talwana H, Badji A, Kyalo M, Nasser Y, Gemenet D, Kitavi M, Mwanga ROM (2020) Genome-wide association study identified candidate genes controlling continuous storage root formation and bulking in hexaploid sweetpotato. Bmc Plant Biol 20(1). https://doi.org/10.1186/s12870-019-2217-9

Batista LG, Mello VH, Souza AP, Margarido GRA (2022) Genomic prediction with allele dosage information in highly polyploid species. Theor Appl Genet 135(2):723–739. https://doi.org/10.1007/s00122-021-03994-w

Cervantes-Flores JC, Yencho GC, Kriegner A, Pecota KV, Faulk MA, Mwanga ROM, Sosinski BR (2008) Development of a genetic linkage map and identification of homologous linkage groups in sweetpotato using multiple-dose AFLP markers. Mol Breeding 21(4):511–532. https://doi.org/10.1007/s11032-007-9150-6

Chang KY, Lo HF, Lai YC, Yao PJ, Lin KH, Hwang SY (2009) Identification of quantitative trait loci associated with yield-related traits in sweet potato (*Ipomoea batatas*). Bot Stud 50:43–55

Clark LV, Lipka AE, Sacks EJ (2019) polyRAD: genotype calling with uncertainty from sequencing data in polyploids and diploids. G3 (Bethesda) 9(3):663–673. https://doi.org/10.1534/g3.118.200913

Connolly AG, Godwin ID, Cooper M, DeLacy IH (1994) Interpretation of randomly amplified polymorphic DNA marker data for fingerprinting sweetpotato (*Ipomoea batatas* L.) genotypes. Theor Appl Genet 88:332–336. https://doi.org/10.1007/BF00223641

Cooke DP, Wedge DC, Lunter G (2021) A unified haplotype-based method for accurate and comprehensive variant calling. Nat Biotechnol 39(7):885–892. https://doi.org/10.1038/s41587-021-00861-3

da Silva PG, Gemenet DC, Mollinari M, Olukolu BA, Wood JC, Diaz F, Mosquera V, Gruneberg WJ, Khan A, Buell CR, Yencho GC, Zeng ZB (2020) Multiple QTL mapping in autopolyploids: a random-effect model approach with application in a hexaploid sweetpotato full-sib population. Genetics 215(3):579–595. https://doi.org/10.1534/genetics.120.303080

Elameen A, Larsen A, Fjellheim SS, Sundheim L, Msolla S, Masumba E, Rognli OA (2011) Phenotypic diversity of plant morphological and root descriptor traits within a sweet potato, *Ipomoea batatas* (L.) Lam., germplasm collection from Tanzania. Genet Resour Crop Evol 58:397–407. https://doi.org/10.1007/s10722-010-9585-1

Elshire RJ, Glaubitz JC, Sun Q, Poland JA, Kawamoto K, Buckler ES, Mitchell SE (2011) A robust, simple genotyping-by-sequencing (GBS) approach for high diversity species. PLoS ONE 6(5):e19379. https://doi.org/10.1371/journal.pone.0019379

Fajardo S, La Bonte DR, Jarret RL (2002) Identifying and selecting for genetic diversity in Papua New Guinea sweetpotato (L.) Lam. germplasm collected as botanical seed. Genet Resour Crop Evol 49(5):463–470. https://doi.org/10.1023/A:1020955911675

Gemenet DC, Lindqvist-Kreuze H, De Boeck B, Pereira GD, Mollinari M, Zeng ZB, Yencho GC, Campos H (2020b) Sequencing depth and genotype quality: accuracy and breeding operation considerations for genomic selection applications in autopolyploid crops. Theor Appl Genet 133(12):3345–3363. https://doi.org/10.1007/s00122-020-03673-2

Gemenet DC, Kitavi MN, David M, Ndege D, Ssali RT, Swanckaert J (2020a) Development of diagnostic SNP markers for quality assurance and control in sweetpotato [*Ipomoea batatas* (L.) Lam.] breeding programs. Plos One 15(5):e0233828. https://doi.org/10.1371/journal.pone.0232173

Gemenet DC, Pereira GD, De Boeck B, Wood JC, Mollinari M, Olukolu BA, Diaz F, Mosquera V, Ssali RT, David M, Kitavi MN, Burgos G, Zum Felde T, Ghislain M, Carey E, Swanckaert J, Coin LJM, Fei Z, Hamilton JP, Yada B, Yencho GC, Zeng ZB, Mwanga ROM, Khan A, Gruneberg WJ, Buell CR (2020c) Quantitative trait loci and differential gene expression analyses reveal the genetic basis for negatively associated β-carotene and starch content in hexaploid sweetpotato [*Ipomoea batatas* (L.) Lam.]. Theoret Appl Genet 133(1):23–36. https://doi.org/10.1007/s00122-019-03437-7

Gerard D, Ferrao LFV, Garcia AAF, Stephens M (2018) Genotyping polyploids from messy sequencing data. Genetics 210(3):789–807. https://doi.org/10.1534/genetics.118.301468

Gichuki ST, Berenyi M, Zhang DP, Hermann M, Schmidt J, Glössl J, Burg K (2003) Genetic diversity in sweetpotato [*Ipomoea batatas* (L.) Lam.] in relationship to geographic sources as assessed with RAPD markers. Genet Resour Crop Evol 50(4):429–437. https://doi.org/10.1023/A:1023998522845

Grodzicker T, Williams J, Sharp P (1974) Physical mapping of temperature-sensitive mutations of adenovirus.

Cold Spring Harbor Symp Quant Biol 34:439–446. https://doi.org/10.1101/sqb.1974.039.01.056

Haque E, Yamamoto E, Shirasawa K, Tabuchi H, Yoon UH, Isobe S, Tanaka M (2020b) Genetic analyses of anthocyanin content using polyploid GWAS followed by QTL detection in the sweetpotato (*Ipomoea batatas* L.) storage root. Plant Root 14:11–21. https://doi.org/10.3117/plantroot.14.11

Haque E, Tabuchi H, Monden Y, Suematsu K, Shirasawa K, Isobe S, Tanaka M (2020a) QTL analysis and GWAS of agronomic traits in sweetpotato (*Ipomoea batatas* L.) using genome wide SNPs. Breed Sci 70(3):283–291. https://doi.org/10.1270/jsbbs.19099

Huang JC, Sun M (2000) Genetic diversity and relationships of sweetpotato and its wild relatives in *Ipomoea* series Batatas (Convolvulaceae) as revealed by inter-simple sequence repeat (ISSR) and restriction analysis of chloroplast DNA. Theor Appl Genet 100:1050–1060. https://doi.org/10.1007/s001220051386

Huang J, Corke H, Sun M (2002) Highly polymorphic AFLP markers as a complementary tool to ITS sequences in assessing genetic diversity and phylogenetic relationships of sweetpotato (*Ipomoea batatas* (L.) Lam.) and its wild relatives. Genet Resour Crop Evol 49:541–550. https://doi.org/10.1023/A:1021290927362

Iwanaga M (1988) Use of wild germplasm for sweet potato breeding. In: Gregory P (ed) Exploration, maintenance, and utilization of sweet potato genetic resources. Report of the 1st sweet potato planning conference 1987, International Potato Center (CIP), Lima, Peru

Jarret RL, Gawe N, Whittemore A (1992) Phylogenetic Relationships of the Sweetpotato [*Ipomoea batatas* (L.) Lam.]. J Am Soc Hort Sci 117(4):633–637. https://doi.org/10.21273/JASHS.117.4.633

John H, Birnstiel M, Jones K (1969) RNA-DNA hybrids at the cytological level. Nature 223:582–587. https://doi.org/10.1038/223582a0

Jones A (1967) Should Nishiyama's K123 (*Ipomoea trifida*) be designated I. batatas? Econ Bot 21:163–166. http://www.jstor.org/stable/4252862

Kim JH, Chung IK, Kim KM (2017) Construction of a genetic map using EST-SSR markers and QTL analysis of major agronomic characters in hexaploid sweet potato (*Ipomoea batatas* (L.) Lam). Plos One 12: e0185073–9. https://doi.org/10.1371/journal.pone.0185073

Kriegner A, Cervantes JC, Burg K, Mwanga ROM, Zhang D (2003) A genetic linkage map of sweetpotato [*Ipomoea batatas* (L.) Lam.] based on AFLP markers. Mol Breed 11:169–185. https://doi.org/10.1023/A:1022870917230

Lee KJ, Lee GA, Lee JR, Sebastin R, Shin MJ, Cho GT, Hyun D (2019) Genetic diversity of sweet potato (L. Lam) germplasms collected worldwide using chloroplast SSR markers. Agron-Basel 9(11). https://doi.org/10.3390/agronomy9110752

Lewontin RC, Hubby JL (1966) A molecular approach to the study of genic heterozygosity in natural

populations. II. Amount of variation and degree of heterozygosity in natural populations of *Drosophila pseudoobscura*. Genetics 54:595–609. https://doi.org/10.1093/genetics/54.2.595

Li AX, Liu Q, Wang Q, Zhang L, Zhai H, Liu S (2010) Construction of molecular linkage maps using srap markers in sweetpotato. Acta Agron Sin 36:1286–1295. https://doi.org/10.1016/S1875-2780(09)60065-1

Ma ZM, Gao WC, Liu LF, Liu MH, Zhao N, Han MK, Wang Z, Jiao WJ, Gao ZY, Hu YY, Liu QC (2020) Identification of QTL for resistance to root rot in sweetpotato ((L.) Lam) with SSR linkage maps. Bmc Genom 21(1). https://doi.org/10.1186/s12864-020-06775-9

Maquia I, Muocha I, Naico A, Martins N, Gouveia M, Andrade I, Goulao LF, Ribeiro AI (2013) Molecular, morphological and agronomic characterization of the sweetpotato (*Ipomoea batatas* L.) germplasm collection from Mozambique: genotypeselection for drought prone regions. South Afr J Bot 88:142–151. https://doi.org/10.1016/j.sajb.2013.07.008

McDonald JA, Austin DF (1990) Changes and additions in *Ipomoea* section Batatas (Convolvulaceae). Brittonia 42:116–120. https://doi.org/10.2307/2807625

Meng YS, Zheng CX, Li H, Li AX, Zhai H, Wang QM, He SZ, Zhao N, Zhang H, Gao SP, Liu QC (2021) Development of a high-density SSR genetic linkage map in sweet potato. Crop J 9(6):1367–1374. https://doi.org/10.1016/j.cj.2021.01.003

Meng YS, Zhao N, Li H, Zhai H, He SZ, Liu QC (2018) SSR fingerprinting of 203 sweetpotato (*Ipomoea batatas* (L.) Lam.) varieties. J Integr Agric 17(1):86–93. https://doi.org/10.1016/S2095-3119(17)61687-3

Moeinzadeh MH, Yang J, Muzychenko E, Gallone G, Heller D, Reinert K, Haas S, Vingron M (2020) Ranbow: a fast and accurate method for polyploid haplotype reconstruction. PLoS Comput Biol 16(5):e1007843. https://doi.org/10.1371/journal.pcbi.1007843

Mollinari M, Olukolu BA, Pereira GDS., Khan A, Gemenet D, Yencho GC, Zeng ZB (2020) Unraveling the hexaploid sweetpotato inheritance using ultra-dense multilocus mapping. G3 (Bethesda) 10(1):281–292. https://doi.org/10.1534/g3.119.400620

Munoz-Rodriguez P, Wells T, Wood JRI, Carruthers T, Anglin NL, Jarret RL, Scotland RW (2022) Discovery and characterization of sweetpotato's closest tetraploid relative. New Phytol 234(4):1185–1194. https://doi.org/10.1111/nph.17991

Naidoo SIM, Laurie SM, Amelework AB, Shimelis H, Laing M (2022) Selection of sweetpotato parental genotypes using simple sequence repeat markers. Plants-Basel 11(14). https://doi.org/10.3390/plants11141802

Obata N, Tabuchi HT, Kurihara MK, Yamamoto EY, Shirasawa KS, Monden YM (2022) Mapping of nematode resistance in hexaploid sweetpotato using a next-generation sequencing-based association study. Front Plant Sci 13. https://doi.org/10.3389/fpls.2022.858747

Okada Y, Monden Y, Nokihara K, Shirasawa K, Isobe S, Tahara M (2019) Genome-wide association studies (GWAS) for yield and weevil resistance in sweet potato (*Ipomoea batatas* (L.) Lam). Plant Cell Rep 38(11):1383–1392. https://doi.org/10.1007/s00299-019-02445-7

Oloka BM, da Silva PG, Amankwaah VA, Mollinari M, Pecota KV, Yada B, Olukolu BA, Zeng ZB, Yencho GC (2021) Discovery of a major QTL for root-knot nematode (*Meloidogyne incognita*) resistance in cultivated sweetpotato (*Ipomoea batatas*). Theor Appl Genet 134(7):1945–1955. https://doi.org/10.1007/s00122-021-03797-z

Orjeda G, Freyre R, Iwanaga M (1990) Production of 2n pollen in diploid *Ipomoea trifida*, a putative wild ancestor of sweet potato. J Hered 81:462–467. https://doi.org/10.1093/oxfordjournals.jhered.a111026

Pardue ML, Gall JG (1969) Molecular hybridization of radioactive RNA to the DNA of cytological preparations. Proc Natl Acad Sci U S A 64:600–604. https://doi.org/10.1073/pnas.64.2.600

Peterson BK, Weber JN, Kay EH, Fisher HS, Hoekstra HE (2012) Double digest RADseq: an inexpensive method for de novo SNP discovery and genotyping in model and non-model species. PLoS ONE 7(5):e37135. https://doi.org/10.1371/journal.pone.0037135

Pinkel D, Segraves R, Sudar D, Clark S, Poole I, Kowbel D, Collins C, Kuo WL, Chen C, Zhai Y, Dairkee SH, Ljung BM, Gray JW, Albertson DG (1998) High resolution analysis of DNA copy number variation using comparative genomic hybridization to microarrays. Nat Genet 20:207–211

Reyes LM, Collins WW (1992) Genetic control of seven enzyme systems in *Ipomoea* species. J Am Soc Hort Sci 117(6):1000–1005. https://doi.org/10.21273/JASHS.117.6.1000

Ronaghi M, Karamohamed S, Pettersson B, Uhlén M, Nyrén P (1996) Real-time DNA sequencing using detection of pyrophosphate release. Anal Biochem 242(1):84–89. https://doi.org/10.1006/abio.1996.0432

Roullier C, Duputié A, Wennekes P, Benoit L, Bringas VMF., Rossel G, Tay D, McKey D, Lebot V (2013) Disentangling the Origins of Cultivated Sweet Potato ((L.) Lam.). Plos One 8(5). https://doi.org/10.1371/journal.pone.0062707

Sagredo B, Hinrichsen P, López H, Cubillos A, Muñoz C (1998) Genetic variation of sweet potatoes (*Ipomoea batatas* L) cultivated in Chile determined by RAPDs. Euphytica 101(2):193–198. https://doi.org/10.1023/A:1018301009296

Saiki RK, Scharf S, Faloona F, Mullis KB, Horn GT, Erlich HA, Arnheim N (1985) Enzymatic amplification of beta-globin genomic sequences and restriction site analysis for diagnosis of sickle cell anemia. Science 230(4732):1350–1354. https://doi.org/10.1126/science.2999980

Samiyarsih S, Fitrianto N, Azizah E, Herawati W, Rochmatino (2020) Anatomical profile and genetic variability of sweet potato (*Ipomoea batatas*) cultivars in Banyumas, Central Java, based on RAPD markers. Biodiversitas 21(4):1755–1766. https://doi.org/10.13057/biodiv/d210460

Schlotterer C (2004) The evolution of molecular markers–just a matter of fashion? Nat Rev Genet 5(1):63–69. https://doi.org/10.1038/nrg1249

Schmitz Carley CA, Coombs JJ, Douches DS, Bethke PC, Palta JP, Novy RG, Endelman JB (2017) Automated tetraploid genotype calling by hierarchical clustering. Theor Appl Genet 130(4):717–726. https://doi.org/10.1007/s00122-016-2845-5

Serang O, Mollinari M, Garcia AA (2012) Efficient exact maximum a posteriori computation for bayesian SNP genotyping in polyploids. PLoS ONE 7(2):e30906. https://doi.org/10.1371/journal.pone.0030906

Shirasawa K, Tanaka M, Takahata Y, Ma DF, Cao QH, Liu QC, Zhai H, Kwak SS, Jeong JC, Yoon UH, Lee HU, Hirakawa H, Isobe S (2017) A high-density SNP genetic map consisting of a complete set of homologous groups in autohexaploid sweetpotato. Sci Rep 7:44207. https://doi.org/10.1038/srep44207

Slonecki TJ, Rutter WB, Olukolu BA, Yencho GC, Jackson DM, Wadl PA (2023) Genetic diversity, population structure, and selection of breeder germplasm subsets from the USDA sweetpotato collection. Front Plant Sci 13:1022555. https://doi.org/10.3389/fpls.2022.1022555

Soegiant A, Ardiarini NR, Sugiharto AN (2011) Genetic Diversity of Sweet Potato (*Ipomoea batatas* L.) in East Java, Indonesia. J Agric Food Tech 1(9):179–183

Southern EM (1975) Detection of specific sequences among DNA fragments separated by gel electrophoresis. J Mol Biol 98(3):503–517. https://doi.org/10.1016/S0022-2836(75)80083-0

Sun X, Liu D, Zhang X, Li W, Liu H, Hong W, Jiang C, Guan N, Ma C, Zeng H, Xu C, Song J, Huang L, Wang C, Shi J, Wang R, Zheng X, Lu C, Wang X, Zheng H (2013) SLAF-seq: an efficient method of large-scale de novo SNP discovery and genotyping using high-throughput sequencing. PLoS ONE 8(3):e58700. https://doi.org/10.1371/journal.pone.0058700

Sun C, Dong Z, Zhao L, Ren Y, Zhang N, Chen F (2020) The Wheat 660K SNP array demonstrates great potential for marker-assisted selection in polyploid wheat. Plant Biotechnol J 18(6):1354–1360. https://doi.org/10.1111/pbi.13361

Taniguti CH, Gesteira GS, Lau J, Pereira GS, Zeng ZB, Byrne D, Riera-Lizarazu O, Mollinari M (2022) VIEWpoly: a visualization tool to integrate and explore results of polyploid genetic analysis. J Open Sour Softw 7(74):4242. https://doi.org/10.21105/joss.04242

Thompson PG, Hong LL, Ukoskit K, Zhu Z (1997) Genetic linkage of randomly amplified polymorphic DNA(RAPD) markers in sweetpotato. J Am Soc Hort Sci 122:79–82. https://doi.org/10.21273/JASHS.122.1.79

Tiselius AA (1937) New apparatus for electrophoretic analysis of colloidal mixtures. Trans Faraday Soc 33:524–531. https://doi.org/10.1039/tf9373300524

Veasey EA, Borges A, Rosa MS, Queiroz-Silva JR, Bressan EDA., Peroni N (2008) Genetic diversity in Brazilian sweet potato (*Ipomoea batatas* (L.) Lam., Solanales, Convolvulaceae) landraces assessed with microsatellite markers. Genet Molecul Biol 31(3):725–733. https://doi.org/10.1590/S1415-47572008000400020

Vos P, Hogers R, Bleeker M, Reijans M, van de Lee T, Hornes M, Frijters A, Pot J, Peleman J, Kuiper M et al (1995) AFLP: a new technique for DNA fingerprinting. Nucleic Acids Res 23(21):4407–4414. https://doi.org/10.1093/nar/23.21.4407

Wadl PA, Olukolu BA, Branham SE, Jarret RL, Yencho GC, Jackson DM (2018) Genetic diversity and population structure of the USDA sweetpotato germplasm collections using GBSpoly. Front Plant Sci 9:1166. https://doi.org/10.3389/fpls.2018.01166

Wang ZY, Fang BP, Chen JY, Zhang XJ, Luo ZX, Huang LF, Chen XL, Li YJ (2010) Assembly and characterization of root transcriptome using Illumina paired-end sequencing and development of cSSR markers in sweetpotato. BMC Genomics 11:1–14. https://doi.org/10.1186/1471-2164-11-726

Wang ZY, Li J, Luo ZX, Huang LF, Chen XL, Fang BP, Li YJ, Chen JY, Zhang XJ (2011) Characterization and development of EST-derived SSR markers in cultivated sweetpotato. BMC Plant Biol 11:1–9. https://doi.org/10.1186/1471-2229-11-139

Williams JG, Kubelik AR, Livak KJ, Rafalski JA, Tingey SV (1990) DNA polymorphisms amplified by arbitrary primers are useful as genetic markers. Nucleic Acids Res 18(22):6531–6535. https://doi.org/10.1093/nar/18.22.6531

Wu S, Lau KH, Cao Q, Hamilton JP, Sun H, Zhou C, Eserman L, Gemenet DC, Olukolu BA, Wang H, Crisovan E, Godden GT, Jiao C, Wang X, Kitavi M, Manrique-Carpintero N, Vaillancourt B, Wiegert-Rininger K, Yang X, Bao K, Schaff J, Kreuze J, Gruneberg W, Khan A, Ghislain M, Ma D, Jiang J, Mwanga ROM, Leebens-Mack J, Coin LJM, Yencho GC, Buell CR, Fei Z (2018) Genome sequences of two diploid wild relatives of cultivated sweetpotato reveal targets for genetic improvement. Nat Commun 9(1):4580. https://doi.org/10.1038/s41467-018-06983-8

Xiao S, Dai X, Zhao L, Zhou Z, Zhao L, Xu P, Gao B, Zhang A, Zhao D, Yuan R, Wang Y, Wang J, Li Q, Cao Q (2023) Resequencing of sweetpotato germplasm resources reveals key loci associated with multiple agronomic traits. Hortic Res 10(1):uhac234. https://doi.org/10.1093/hr/uhac234

Yada B, Tukamuhabwa P, Wajala B, Kim DJ, Skiton RA, Alajo BA, Mwanga ROM (2010) Characterization of Uganda sweet potato germplasm using fluorescence repeat markers. Horticult Sci 45(2):225–230. https://doi.org/10.21273/HORTSCI.45.2.225

Yamakawa H, Haque E, Tanaka M, Takagi H, Asano K, Shimosaka E, Akai K, Okamoto S, Katayama K, Tamiya S (2021) Polyploid QTL-seq towards rapid development of tightly linked DNA markers for potato and sweetpotato breeding through whole-genome resequencing. Plant Biotechnol J 19(10):2040–2051. https://doi.org/10.1111/pbi.13633

Yamamoto E, Shirasawa K, Kimura T, Monden Y, Tanaka M, Isobe S (2020) Genetic mapping in autohexaploid sweet potato with low-coverage NGS-based genotyping data. G3 10:2661–2670

Yan M, Li M, Wang Y, Wang X, Moeinzadeh MH, Quispe-Huamanquispe DG, Fan W, Fang Y, Wang Y, Nie H, Wang Z, Tanaka A, Heider B, Kreuze JF, Gheysen G, Wang H, Vingron M, Bock R, Yang J (2024) Haplotype-based phylogenetic analysis and population genomics uncover the origin and domestication of sweetpotato. Mol Plant 17(2):277–296. https://doi.org/10.1016/j.molp.2023.12.019

Yan H, Ma M, Ahmad MQ, Arisha MH, Tang W, Li C, Zhang Y, Kou M, Wang X, Gao R, Song W, Li Z, Li Q (2022) High-density single nucleotide polymorphisms genetic map construction and quantitative trait locus mapping of color-related traits of purple sweet potato [*Ipomoea batatas* (L.) Lam.]. Front Plant Sci 12:797041. https://doi.org/10.3389/fpls.2021.797041

Yu XX, Zhao N, Li H, Jie Q, Zhai H, He SZ, Li Q, Liu QC (2014) Identification of QTLs for starch content in sweetpotato (*Ipomoea batatas* (L.) Lam.). J Integr Agric 13(2):310–315. https://doi.org/10.1016/S2095-3119(13)60357-3

Zhang D, Ghislain M, Huamán Z, Golmirzaie A, Hijmans R (1998) RAPD variation in sweetpotato (*Ipomoea batatas* (L.) Lam) cultivars from South America and Papua New Guinea. Genet Resour Crop Evol 45:271–277. https://doi.org/10.1023/A:1008642707580

Zhang K, Wu Z, Tang D, Lv C, Luo K, Zhao Y, Liu X, Huang Y, Wang J (2016) Development and identification of SSR markers associated with starch properties and β-carotene content in the storage root of sweet potato (*Ipomoea batatas* L.). Front Plant Sci 7:1–21. https://doi.org/10.3389/fpls.2016.00223

Zhang DP, Cervantes J, Huamán Z, Carey E, Ghislain M (2000) Assessing genetic diversity of sweet potato (*Ipomoea batatas* (L.) Lam.) cultivars from tropical America using AFLP. Genet Resour Crop Evol 47(6):659–665. https://doi.org/10.1023/A:1026520507223

Zhang DP, Rossel G, Kriegner A, Hijmans R (2004) AFLP assessment of diversity in sweetpotato from Latin America and the Pacific region: Its implications on the dispersal of the crop. Genet Resour Crop Evol 51(2):115–120. https://doi.org/10.1023/B:GRES.0000020853.04508.a0

Zhao N, Yu XX, Jie Q, Li H, Li H, Hu J, Zhai H, He SZ, Liu QC (2013a) A genetic linkage map based on AFLP and SSR markers and mapping of QTL for dry-matter content in sweetpotato. Mol Breeding 32(4):807–820. https://doi.org/10.1007/s11032-013-9908-y

Zhao N, Zhai H, Yu XX, Liu ZS, He SZ, Li Q, Ma DF, Liu QC (2013b) Development of SRAP markers linked to a gene for stem nematode resistance in sweetpotato, (L.) Lam. J Integr Agric 12(3):414–419. https://doi.org/10.1016/S2095-3119(13)60241-5

Zheng CX, Jiang ZC, Meng YS, Yu J, Yang XS, Zhang H, Zhao N, He SZ, Gao SP, Zhai H, Liu QC (2023) Construction of a high-density SSR genetic linkage map and identification of QTL for storage-root yield and dry-matter content in sweetpotato. Crop J 11(3):963–967. https://doi.org/10.1016/j.cj.2022.11.003

Zietkiewicz E, Rafalski A, Labuda D (1994) Genome fingerprinting by simple sequence repeat (SSR)-anchored polymerase chain reaction amplification. Genomics 20(2):176–183. https://doi.org/10.1006/geno.1994.1151

Zych K, Gort G, Maliepaard CA, Jansen RC, Voorrips RE (2019) FitTetra 2.0—improved genotype calling for tetraploids with multiple populations and parental data support. BMC Bioinformatics 20(1):148. https://doi.org/10.1186/s12859-019-2703-y

Genetic Maps in Sweetpotato

5

Gabriel de Siqueira Gesteira,
Guilherme da Silva Pereira, Zhao-Bang Zeng,
and Marcelo Mollinari

Abstract

This chapter highlights the research and efforts that have been done to understand the composition of the genome and the mechanisms underlying the genetic inheritance in sweetpotato, with focus on the cultivated hexaploid sweetpotato. The first part of the chapter focuses on dissecting strategies and methods that have been used to study and understand key factors that affect the genetic behavior in polyploid species, with emphasis on linkage analysis, highlighting the most common types of experimental populations used for genetic mapping, the obtention of genotype information, and the choice of analytical methods to study such populations. The second part of the chapter dives deeper into the knowledge accumulated through the application of traditional methods and the more recent adoption of cutting-edge technologies, combined with state-of-the-art algorithms that were developed specifically for polyploid species, to study and shed a light on the genetic architecture and the mechanisms that drive the genetic transmission in the cultivated hexaploid sweetpotato.

Keywords

Genetic mapping · Linkage analysis · Polyploid · Autopolyploid · Sweetpotato

5.1 Introduction

Genetic linkage mapping is a fundamental tool for understanding the inheritance mechanism in cultivated crops. It is important for identifying and characterizing genomic regions associated with agronomic traits, supporting evolutionary studies, and assisting the assembly of reference genomes. Linkage analysis has been playing an important role in biology and, more specifically, in genetic studies. It started with the investigations of Thomas Hunt Morgan and his team conducting pioneering experiments with *Drosophila melanogaster* (fruit flies) that led to the discovery of gene linkage and the concept of genetic recombination (Morgan 1911). Further, he and his team, including his doctoral student Alfred Sturtevant, noticed that the frequency of certain traits being inherited together varied in a way

G. de Siqueira Gesteira · Z.-B. Zeng ·
M. Mollinari (✉)
Department of Horticultural Science, Bioinformatics Research Center, North Carolina State University, Raleigh, NC 27695, USA
e-mail: mmollin@ncsu.edu

G. de Siqueira Gesteira
e-mail: gdesiqu@ncsu.edu

Z.-B Zeng
e-mail: szeng@ncsu.edu

G. da Silva Pereira
Department of Agronomy, Federal University of Viçosa, Viçosa, MG, Brazil
e-mail: g.pereira@ufv.br

G. C. Yencho et al. (eds.), *The Sweetpotato Genome*, Compendium of Plant Genomes,
https://doi.org/10.1007/978-3-031-65003-1_5

that could be correlated with their relative positions on the chromosome.

Building on these observations, Sturtevant, constructed the first genetic map, demonstrating the linear arrangement of genes on a chromosome (Sturtevant 1913). This work fundamentally showed that the closer two genes were to each other on a chromosome, the less likely they were to be separated during genetic recombination, leading to the concept of linkage maps. Their contributions were facilitated by key characteristics of the *Drosophila* model, including fast and easy manipulation of experimental populations (small and short-life organism that allows inbreeding) and a small number of large chromosomes ($2n = 2x = 8$) that carry the genetic information of relatively simple morphological traits, such as eye color, wing format, and wing size. However, other organisms proved to be more complex than *Drosophila*, and it soon became clear that, although the basic concepts of linkage remained the same, extensions and modifications to the initial linkage analysis were necessary.

In the initial works on genetic linkage maps, a widely accepted assumption was that the linkage phase, i.e., the arrangement or orientation of alleles in the homologous chromosomes, was known for the parents in the studied population. Knowing the parental linkage phases simplifies the task of distinguishing between recombinant and non-recombinant individuals. This applies particularly to the most frequently utilized crossing structures, including F_2 generations, backcrosses, and Recombinant Inbred Lines (RILs) (Mollinari et al. 2009). In that context, the process of building a genetic map can be summarized into three steps: (1) calculating the recombination fractions for all pairs of markers; (2) grouping markers into linkage groups; (3) ordering markers within linkage groups. While these three steps remain valid and de novo maps can still be constructed by applying them, recent advancements in sequencing technologies have

facilitated the use of prior genomic information to group and order genetic markers more effectively. For the first step, calculating the recombination fractions is contingent upon the crossing structure of the mapping population. These calculations are detailed extensively in (Liu 1998), and usually follow the general formula:

$$rf_{ij} = \frac{R}{R + NR}$$

In this formula, rf_{ij} is the estimated recombination fraction between loci i and j, R is the number of observed recombinant offspring, and NR is the number of observed non-recombinant offspring for both loci. The calculated recombination fractions can be readily converted into genetic distances, under certain assumptions, by employing one of the following mapping functions (Morgan 1917; Haldane 1919; Kosambi 1943):

$$d_{ij} = r_{ij}$$

$$d_{ij} = -\frac{1}{2}\ln(1 - 2r_{ij})$$

$$d_{ij} = \frac{1}{4}\ln\left[\frac{1 + 2r_{ij}}{1 - 2r_{ij}}\right]$$

where d_{ij} is the genetic distance between loci i and j in centimorgan (cM), and r_{ij} is the recombination fraction between loci i and j.

Estimating the recombination fraction can become complex when the dataset has missing observations or when deriving the population from homozygous lines proves impractical. Under these circumstances, it becomes essential not only to tackle the missing data, but also to accurately infer the linkage phase configurations to ensure the integrity of the genetic map construction process. In such cases, more sophisticated methodologies are required, such as the general maximum-likelihood-based algorithm for simultaneously estimating linkage and linkage phases for markers with varying degrees of missingness, presented by Wu et al. (2002a, b) and implemented in the R package OneMap (Margarido et al. 2007).

This scenario becomes significantly more complex when dealing with organisms that possess multiple copies of their entire chromosome set, known as **polyploids**. Polyploid organisms are classified into *autopolyploids*, where the multiple chromosome sets originate from the same species, and *allopolyploids*, where the sets come from different species. In most instances, allopolyploids demonstrate segregation patterns similar to diploids, primarily because their homologous chromosomes usually form bivalents within each sub-genome (preferential pairing). On the other hand, autopolyploids often display either random bivalent formation or the formation of multivalents during meiosis, leading to more complex polysomic segregation patterns (Sybenga 1975; Soltis and Soltis 1993; Osborn et al. 2003; Mollinari and Garcia 2019).

In this chapter, we integrate the foundational principles of genetic linkage mapping and the specific complexities of the autohexaploid genome of sweetpotato (*Ipomoea batatas* (L.) Lam., $2n = 6x = 90$). We examine the challenges and specialized methodologies required to build genetic linkage maps in a polyploid context, highlighting the differences from simpler diploid genetic models. First, we outline the differences in types of experimental populations commonly used in genetic mapping, such as backcrosses, F_2, and RILs. We discuss the limitations that often restrict their application in polyploid mapping, specifically focusing on sweetpotato linkage mapping. Subsequently, we examine the inheritance patterns in autopolyploid organisms and then examine genotyping techniques employed in constructing genetic maps for autopolyploids, providing a historical perspective on the use of these techniques in the genetic mapping of autopolyploids. Finally, we concentrate on the history of sweetpotato genetic mapping and present our group's contributions to the current state-of-the-art in sweetpotato genetic maps.

5.2 Types of Experimental Populations Used in Genetic Mapping

Experimental populations are essential in various genetic mapping studies, including linkage analysis, quantitative trait loci (QTL) mapping, and candidate gene identification. The available choices and definitions of optimal experimental populations for a given species depend on its particularities and the objectives of the experiment (Doerge et al. 1997; Lynch and Walsh 1998; Doerge 2002). This can include their reproductive mechanisms, genetic diversity, feasibility of controlled crosses, and availability of resources. Such populations are designed to achieve specific goals in genetic mapping studies or to meet predefined objectives in a breeding program.

In several diploid species, the possibility of selfing individuals and obtaining inbred lines has been utilized as an advantageous tool for breeding and genetic mapping studies. In major diploid crops, such as maize, soybean, and rice, experimental populations for genetic mapping are usually derived from crosses between homozygous or inbred lines. Depending on the predominant reproductive mechanism of the target species, these inbred lines can be readily available, obtained by repeated self-fertilization of heterozygous material, or by double-haploidization techniques. Once two inbred lines are crossed, all first-generation (F_1) individuals will be identical hybrids, while individuals formed in later self-pollination stages will segregate and increase homozygosity accordingly. Several experimental populations can be obtained by backcrossing or selfing strategies using the F_1 individuals and founder parents (Lynch and Walsh 1998).

In experimental populations derived from inbred lines, the founder individuals will present homozygous genotypes for all loci. Consequently, the offspring will be composed of the known founder genotypes recombined according to the genetic distances between loci. This is valid for all inbred-based designs, such as RILs, backcrosses, Nested Association Mapping (NAMs), F_2, and others. In such cases, the only variables to be estimated are the recombination frequency rates between markers in the genome (Liu 1998; Broman and Sen 2009). Hence, the phasing procedure, i.e., assessing the haplotype composition of individuals in the population, becomes trivial since the founder haplotypes are known by design.

However, many crop species do not tolerate self-pollination or the obtention of inbred lines. Several biological mechanisms might be in place to circumvent inbreeding, which include dioecy, chasmogamy, self-incompatibilities, spatial and temporal barriers, and others (Soltis and Soltis 2012). When one or more of these mechanisms are present, outcrossing might become the major form of reproduction, and the genetic structure of the populations can be associated with high levels of heterozygosity. Experimental populations can be obtained in such cases, but since founder genotypes will be composed of several heterozygous loci, the F_1 population will present genetic segregation and the assortment of alleles in the homologous chromosomes is not defined by design (i.e., parental haplotypes are unknown).

Dealing with outcrossing species is usually associated with additional layers of complexity that sit on top of well-known practical challenges, such as making controlled crosses between individuals that present incompatibilities, unsynchronized reproductive maturity, production of a small number of seeds, and others. From a genetic perspective, the complexity often lies in the unknown linkage phase of heterozygous genotypes. These heterozygous loci complicate the analysis, as it becomes necessary not only to estimate the recombination frequencies but also to determine the genetic linkage phases between loci. Since one depends on the other, this dual estimation task significantly contributes to the complexity of genetic analysis (Wu et al. 2002a, b). It can be exponentially challenging in other scenarios, such as complex crossing schemes (i.e., diallel crosses and breeding designs), especially with higher ploidy levels (Serang et al. 2012; Mollinari and Garcia 2019).

Similarly as observed for outcrossing diploid species, obtaining homozygous lines in most autopolyploid crops, such as sweetpotato, becomes impractical. While a high level of homozygosity can be achieved with five to six self-generations in diploid species, this number is much higher in autopolyploids. For instance, in diploids, it is possible to obtain approximately 97% of homozygosity with five self-generations from a heterozygous individual Aa (Fig. 5.1). On the other hand, in autohexaploids, this level

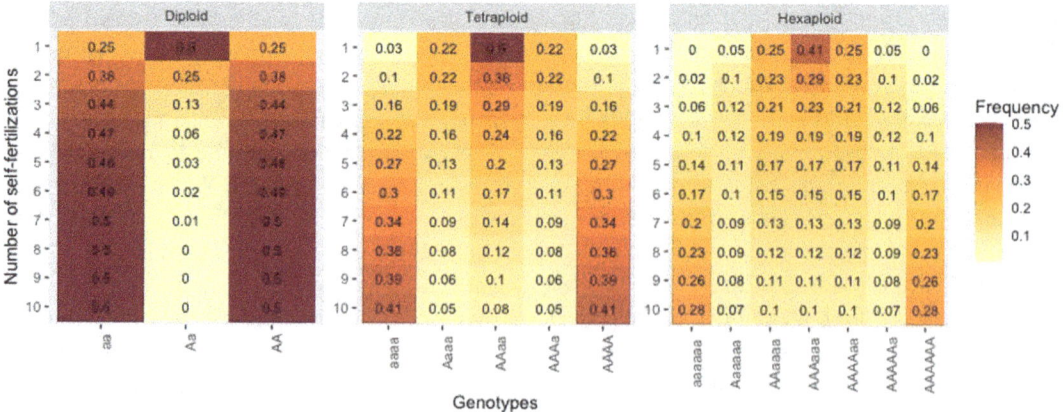

Fig. 5.1 Frequency of genotypic classes through selfing heterozygous in diploid (Aa), tetraploid (AAaa), and hexaploid (AAAaaa) genotypes, respectively. Notice that diploids achieve 97% homozygosity in five self-generations, while in hexaploids, only 56% of the genotypes are homozygous in the same number of self-generations

of homozygosity could only be obtained after 34 self-fertilizations from a heterozygous individual AAAaaa. Despite the impracticality of the long time necessary to obtain that many generations, most autopolyploids present biological mechanisms to prevent inbreeding, such as inbreeding depression, incompatibilities and constraints to self-fertilization, which makes the obtention of homozygous materials much more difficult.

Therefore, polyploid genetic mapping studies have extensively used experimental populations derived from bi-parental (or full-sib) crosses between heterozygous genotypes due to their practical aspects, relatively low resource demand, and genetic properties. Various statistical genetics methods have been developed to capture the Mendelian inheritance and facilitate genetic mapping studies in outcrossing and polyploid species (Wu et al. 1992, 2004; Grattapaglia and Sederoff 1994; Hackett et al. 2001; Luo et al. 2004; Bourke et al. 2018b; Mollinari and Garcia 2019) and examples of studies utilizing full-sib populations in outcrossing and polyploids are vastly available in the literature (Hackett et al. 2001, 2013; Ming 2001; Hackett 2003; Pastina et al. 2011; Margarido et al. 2015; Balsalobre et al. 2017; Ferreira et al. 2019; Deo et al. 2020; Mollinari et al. 2020; da Pereira et al. 2020; Cappai et al. 2020).

In addition to full-sib populations, other experimental population designs have been recently developed and employed in genetic mapping studies to increase genetic diversity and capture different allelic combinations. Diallel crosses, where multiple full-sib populations are obtained by crossing a set of founders in all possible combinations (complete diallel) or partial combinations (partial diallels), have been increasingly used for both diploid (Rosyara et al. 2013; Bink et al. 2014) and polyploid species (Zheng et al., 2021). These experimental populations provide a broader genetic basis for mapping studies and can enhance the detection of QTL associated with complex traits. In addition to their benefits, these populations might be already in place or have the potential to be utilized for different purposes other than discovery

studies, which is especially relevant for breeding programs with limited resources. Similarly to the advancements in studying bi-parental outcrossing and polyploid populations, novel statistical methods have been developed and implemented for analyzing populations resulting from diallel crosses (Amadeu et al. 2021; Zheng et al. 2021).

Other variations of experimental population designs have also been used in genetic mapping studies, including factorial, top-crosses, and poly-crosses, all composed of a combination of multiple full-sib populations where parental haplotypes were recombined in a single generation. These populations can also benefit from tools developed to analyze diallel populations (Amadeu et al. 2021) since their basic unit and genetic structure are essentially the same. Other experimental population structures can involve multiple generations, such as NAMs (Yu et al. 2008; Buckler et al. 1979; Nice et al. 2017; Song et al. 2017) and the Multiparent Advanced Generation Inter-Cross (MAGIC) design (Huang et al. 2015). However, despite their adoption for genetic mapping studies in diploid species where inbred lines are possible (Yu et al. 2008; Buckler et al. 2009), no statistical methods or tools are available to perform genetic mapping studies in populations that involve multiple generations for outcrossing or polyploid species so far.

The ultimate challenge in genetic mapping studies is the inclusion of more complex scenarios and population structures closer to a true breeding population. These can include multiple combinations of various experimental population design in both diploid and polyploid settings, with multiple generations and even populations derived from individuals with mixed ploidy levels. Including such complex structures in genetic mapping studies would constitute the tipping point for most breeding programs, where populations developed for production may also be used for discovery purposes.

Although using breeding populations for genetic mapping studies would be ideal, many other challenges arise in such scenarios due to their complexity. They usually involve multiple

Fig. 5.2 Diagram showing a balanced sweetpotato gamete with three copies of each of the 15 chromosomes

crosses between highly heterozygous genotypes across multiple generations, which can also include a lack of genetic relatedness between individuals. These conditions impose several difficulties to track and account for recombinations, especially when high ploidy levels and multiple generations are involved. Thus, the design and analysis of experimental populations require careful consideration of the ploidy level, allele dosage effects, and statistical methods suitable for the species under study. Furthermore, computational tools and algorithms specifically tailored for polyploid mapping are crucial for accurate genetic analysis in these species.

5.3 Inheritance Patterns in Sweetpotato

The key element in a genetic mapping study is inferring the genome of offspring individuals based on the recombination of the founders' genomes. Addressing this problem in sweetpotato requires careful consideration of its unique meiotic characteristics. Sweetpotato can be classified as a functional autopolyploid. This classification is significant because, despite the ongoing debate in the scientific literature about the origin of its multiple genomes, sweetpotato displays inheritance patterns typical of an autohexaploid. These patterns are characterized by predominant bivalent formation with random chromosome pairing, and the sporadic occurrence of multivalents during meiosis (Mollinari et al. 2020).

In polyploids, the formation of viable gametes usually involves each resulting cell receiving a balanced subset of the organism's multiple chromosome sets. During meiosis, chromosomes undergo pairing and segregation. However, unlike diploids, where each chromosome pair segregates into different gametes, polyploids must manage multiple sets of chromosomes.

The outcome is the production of gametes with a balanced number and combination of chromosomes, which is crucial for the genetic stability and viability of the offspring (Zielinski and Mittelsten Scheid 2012; Soares et al. 2021). In this context, we typically expect gametes to contain half the ploidy level of the total chromosome set. As sweetpotato is an autohexaploid with a basic number of 15 chromosomes ($2n = 6x = 90$), a balanced gamete would ideally contain three copies of each of the 15 homology groups, resulting in a total of 45 chromosomes ($n = 3x = 45$) (Fig. 5.2).

Note that there are 20 possible ways to choose three homologs from a complete set of six, represented by the binomial coefficient $\binom{6}{3} = 20$. When combining two gametes in a simple cross, a staggering 400 genotypes can emerge in a single generation for a specific genome position. This number is 100 times greater than the four genotypes expected in a diploid cross, even though the ploidy level is only three times higher. This example underscores the exponential increase in genetic complexity associated with higher ploidy levels in autopolyploids (Mollinari et al. 2020). Even though there are 400 possible genotypes for a single position in a hexaploid cross, modern molecular techniques, such as Single Nucleotide Polymorphisms (SNPs), predominantly yield biallelic data. This characteristic often leads to a scenario where, depending on which specific homolog exhibits a given biallelic variation, several resulting genotypes are categorized into broader classes rather than identified individually. This grouping produces intricate and complex segregation patterns in the analysis, presenting additional challenges in accurately interpreting and understanding the genetic behavior and inheritance in these species. Figure 5.3 compares diploid and hexaploid crosses when evaluated using two types of markers: a completely informative multiallelic marker, and a biallelic marker. For the diploid case (A) with a biallelic marker, each homolog is assessed either with variants *A* or *a*, and the four possible genotypes are reduced to three genotypic classes where only the two heterozygous combinations are merged to a single class and cannot be distinguished. In the hexaploid context (B) with a biallelic marker, each parent contributes three homologs with variant *A* and three with variant *a* (triple-dose or triplex marker), but now the 400 possible genotypes are merged into seven genotypic classes where genotypes within each class are not distinguishable, which represents a drastic reduction in the informativeness of the marker. For the diploid cross, the expected segregation ratio is the well-known 1:2:1. Conversely, in the hexaploid cross, the segregation ratio becomes a more complex 1:18:99:164:99:18:1, reflecting the increased genetic complexity inherent in polyploid inheritance.

Given this genetic complexity, numerous challenges can arise in dealing with polyploid species. Besides the inherent genetic intricacies, most polyploid species exhibit some level of incompatibility, often imposing constraints on their breeding or mating strategies, as well as on the types of populations that can be created and obtained (Gallais 2003). This aspect is particularly critical for genetic studies, such as linkage and QTL mapping, where specific experimental designs and known genetic segregation patterns are required in the target population. With genetic incompatibilities present, individuals tend to exhibit a high degree of heterozygosity, leading to extensive segregation in the progenies of any viable cross. Consequently, appropriate genetic designs are essential when studying the genetic behavior of polyploid species.

5.4 Assessing Genotypes in Polyploids

From the 1910s through the 1960s, several studies laid the theoretical groundwork for understanding inheritance and genetic linkage analysis in polyploid organisms, with significant contributions from researchers such as Muller (1914), Haldane (1930), De Winton and Haldane (1931), Mather (1935, 1936), Wright (1938), Fisher (1947, 1943), and Elandt-Johnson (1967). The pioneering concepts of employing molecular markers in autopolyploids were introduced in the early 1990s by Sorrells (1992). Wu et al. (1992) suggested using restriction fragment length polymorphism (RFLP) markers present on single homologous chromosomes, referred to as single-dose or simplex markers. These markers, which reflect the concept of **dosage** in polyploids by indicating the number of copies of a particular allele, segregate in a 1:1 ratio in gametes. Therefore, when a simplex marker is unique to one parent, the resulting observed genotypes will also follow a 1:1 segregation pattern. In polyploid crosses, simplex markers are analogous to heterozygous markers in diploid crosses. This similarity allows the use of standard diploid mapping software and methodologies for linkage analysis and genetic map construction in polyploids. The linkage phase between markers can be deduced by evaluating the likelihood of competing models for restriction fragments found on the same (in coupling) or different (in repulsion) homologs. Several genetic maps of sweetpotato have been developed employing simplex markers as a framework (Ukoskit and Thompson 1997; Kriegner et al. 2003; Cervantes-Flores et al. 2008; Ai-xian et al. 2010; Zhao et al. 2013; Monden and Tahara 2017; Shirasawa et al. 2017).

While simplex markers are a useful approach for navigating the complexities of map construction in polyploid genomes, they tend to oversimplify and constrict our understanding of the inheritance process within such genomes. For example, simplex markers facilitate the creation of maps based on individual homologs but do not account for their interactions within their respective homology groups. On the other hand, multi-dose (or multiplex) markers allow the integration of multiple homologs from the same homology group into a comprehensive genetic map. This approach has roots in practical research, as seen in the morphological marker studies by Lawrence (1929) and Fisher and Mather (1940), and gained momentum with its application in sugarcane by Da Silva (1993a, b). Nevertheless, the early molecular marker technologies, such as Random Amplified Polymorphic DNA (RAPD) and Amplified Fragment Length Polymorphism (AFLP), were limited by their density and dominant nature. This scarcity of multiplex markers led to a constrained integration of homolog maps. For example, if we consider a triplex dominant marker in Fig. 5.3, the complex segregation ratio of 1:18:99:164:99:18:1 is condensed to a simplified ratio of 1:399 that only differentiates the aaaa genotype from the remaining genotypes,

which are combined in a single genotypic class (Haldane 1930; Ripol et al. 1999).

Modern sequencing technologies have advanced rapidly in recent years, leading to a massive increase in genomic data available to researchers. The ability to generate large amounts of sequencing data quickly and cost-effectively has revolutionized the genotyping of polyploid species. While traditional genotyping methods rely on limited scalability techniques, modern sequencing technologies have greatly enhanced the ability to identify genetic variations quickly and accurately in large populations. These modern techniques allow for the

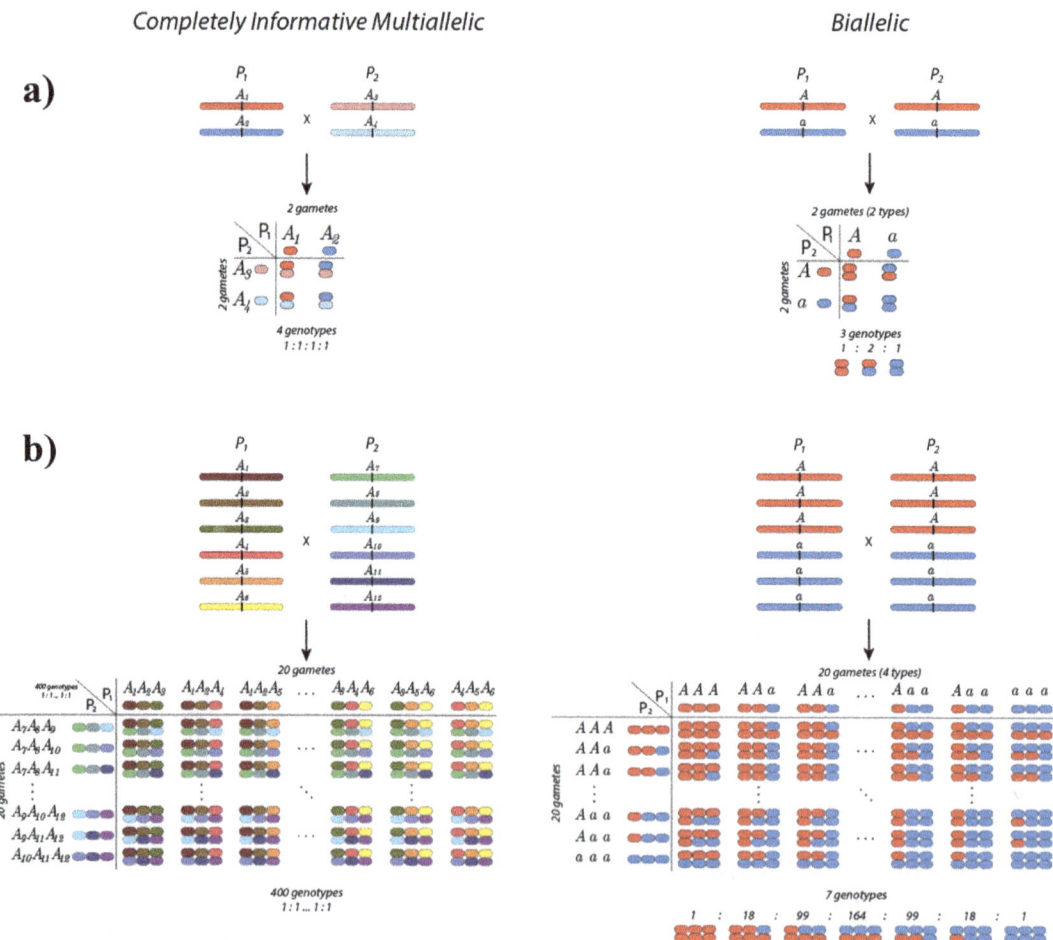

Fig. 5.3 Segregation patterns considering complete informative markers and biallelic markers. **a** in a diploid cross, there are four possible classes in a bi-parental cross and three classes with proportions 1:2:1 when

using biallelic markers. **b** In the hexaploid sweetpotato, 400 possible genotypes are combined in 7 classes when assessed with triplex biallelic markers in both parents. In this case, the expected segregation is 1:18:99:164:99:18:1

assessment of the abundance of reference and alternate alleles, and proper methods for genotype calling are necessary for its use in downstram analyses, especially for outcrossing and polyploid species where heterozygosity plays an important role. Correct genotype calling in polyploid genomes is fundamental to constructing accurate and meaningful genetic maps.

Several methods were proposed and implemented for genotype calling in polyploids. FitTetra and its improved version, FitPoly (Voorrips et al., 2011), implement a classification mixture model weighted by expected frequencies of the genotypic classes in the population. This procedure was implemented to call SNPs in array data, such as Affymetrix Axiom® and Illumina Golden Gate® assays. In these data, the alleles are detected using the fluorescence of two probes using a laser scanning confocal microscope. The reads for the two channels provide a set of ordered pairs of allele intensity for each individual. FitPoly fits a mixture model where parameters are estimated separately for each slice of the population analyzed. As an output, the software provides the probability distribution of the dosage-based genotypic classes for the individuals in the population.

SuperMASSA (Serang et al. 2012) was primarily developed to handle mass-spectrometry-based SNPs. In this case, a Matrix-assisted laser desorption/ionization-time of flight (MALDI-TOF) spectrometer measures the time of flight of the two alleles, each harboring a different size flanking sequence. Sequences with different masses will arrive at different times at the mass spectrometer detection plate, and their abundances are recorded, generating an ordered pair for each SNP. A sequence of ordered pairs is generated in a population of individuals that are classified in terms of their dosage using a Bayesian Network. It also uses the ratio of the two allele channels weighted by the expected genotypic frequencies in the studied population. It implements frequencies based on a Mendelian F_1 segregation model and a Hardy–Weinberg equilibrium model. Although SuperMASSA has been developed to deal with mass spectrometry data, it has been successfully applied in

several studies with different types of genotypic data, including sweetpotato populations that were genotyped with SNPs via Genotyping-by-sequencing (GBS) (Mollinari et al. 2020; Oloka et al. 2021). More recently, Gerard et al. (2018) developed a method to deal with data sets generated by GBS methods and implemented it in the R package *updog*. The package is designed to handle data overdispersion, sequence errors, and genotype biases, elements that are often found in sequence-based genotyping methods. Also, it implements several population models to deal with different assumptions and genotype frequencies. Other methods were proposed, including PolyRAD (Clark et al. 2019) and ClusterCall (Schmitz Carley et al. 2017), but their use is not documented in sweetpotato.

All methods presented in this section provide a dosage-based genotype calling framework for polyploids. For the purpose of genetic mapping, they are often used in populations derived from biparental or interconnected crosses. The next step is to find the relationship between these markers, i.e., linkage analysis, and although it is a well-established procedure in diploids, especially in experimental populations (Lincon et al. 1992; Margarido et al. 2007; Stam 1993), it was not until the recent advancement of high-throughput genomic methods that the full spectrum of genotypes could be incorporated into polyploid linkage analysis. In the next sections, we explore the methodologies for building genetic linkage maps using the different types of molecular data. An overview of the initial strategy of creating separate maps for each parent is presented, then we trace how this approach has informed current methods, preparing us to tackle the unique complexities of polyploidy in genetic mapping studies.

5.5 Initial Polyploid Maps Focusing on Individual Parents

Given the challenges and complexities involved in polyploid genetics, the construction of the first linkage maps in polyploid species started by using methods initially developed for diploid species. Since obtaining inbred-based populations in polyploids is very difficult, several studies have been based on methods that were developed for genetic linkage analysis in outcrossing bi-parental populations, often treating one or both parents as diploids (Grattapaglia and Sederoff 1994; Wu et al. 1992). These methods involve a similar strategy, where simpler segregation cases are used to detect the linkage between markers and estimate the recombination fractions and linkage phase configurations, thus enabling genetic studies in complex and highly heterozygous organisms.

Wu et al. (1992) proposed a polyploid mapping method using single-dose restriction fragments (SDRFs), now commonly called simplex markers, which segregate in a 1:1 ratio in heterozygous plants. They demonstrated this method with hypothetical allopolyploid and autopolyploid species across different ploidy levels to identify SDRFs, detect linkages, and determine genome constitution. The study suggested a minimum population size of 75 to confidently identify simplex markers and detect linkage in coupling phase for both allopolyploids and autopolyploids. However, it noted that meaningful linkages in repulsion were less practical for autopolyploids due to the need for larger populations. Furthermore, the study indicated that the ratio of repulsion to coupling linkages could serve as an indicator of preferential chromosome pairing, which helps differentiate allopolyploids from autopolyploids.

Furthering this approach, the two-way pseudo-testcross strategy was initially devised to analyze linkage in diploid outcrossing species with unknown parental phases (Grattapaglia and Sederoff 1994). Due to its similarities with the approach by Wu et al. (1992), this method has been widely applied to create separate genetic maps for each parent in polyploid species. This method utilizes markers that follow a known Mendelian segregation pattern in the progeny, similar to a test-cross involving one genetically informative parent. For example, consider a diploid biparental cross with a biallelic marker A, where one parent is heterozygous (Aa), and the other is homozygous (aa). In this case, the progeny will show a Mendelian 1:1 segregation of Aa and aa genotypes, given that only the heterozygous parent produces gametes with different alleles. Extending this to two unlinked biallelic markers, A and B, with the first parent heterozygous for both (Aa, Bb) and the second parent homozygous (aa, bb), the progeny will segregate into four genotypic combinations (AaBb, Aabb, aaBb, aabb) in a 1:1:1:1 ratio, according to Mendelian inheritance. If markers A and B are linked, this ratio will alter in proportion to the genetic distance between them. Because only the heterozygous parent provides informative genetic variation, it's the linkage phase configuration of this parent that remains to be estimated. This approach simplifies the process of determining the recombination fraction across all potential linkage phase configurations, whether in coupling or repulsion, and ascertains the most probable configuration.

Beyond their simplicity, a key advantage of using these methods is that the estimators for the recombination fraction are the same for both auto and allopolyploids when using single-dose markers in coupling phase, thus accommodating mapping construction for any ploidy level, including those with intermediate levels of preferential pairing. These methods have seen widespread application in the literature due to their versatility (Porceddu et al. 2002; Chakraborty et al. 2005; Cai et al. 2014; Yang et al. 2016; Vigna et al. 2016). While the approach by Wu et al. (1992) provides a mean to infer the prevalent type of polyploidy, it lacks the more rigorous statistical analysis found in recent advancements such as those by Mollinari et al. (2020) or Bourke et al. (2021).

Although using simplex markers for constructing genetic maps in polyploids offers practical benefits and compatibility with diploid mapping software, it introduces several drawbacks. A significant limitation is the production of separate genetic maps for each parent rather than a single unified map. This occurs because markers segregating in a 1:1 ratio capture recombination events only in the parent they inform about, making the other parent's contribution to those loci invisible. Additionally, this strategy constructs maps on a per-homolog basis rather than per-homology group. This limitation arises because simplex markers located in different homologs (in repulsion) provide minimal information for estimating recombination fractions between them, with feasible estimations only achievable within the same homolog. The lack of information for simplex markers in repulsion becomes more pronounced in higher ploidy levels, which is the case of sweetpotato.

Furthermore, constructing genetic maps for polyploids on a per-homolog basis does not adequately capture the essence of trait expression in those species, as the alleles controlling specific traits are located at analogous positions across all homologs within homology groups. It is, therefore, crucial to treat homologs as part of a homology group rather than as separate entities. Additionally, building a comprehensive map using these strategies would require a density of simplex markers sufficient to ensure complete genome coverage of all homologs for both parents. However, this is not always feasible, sometimes due to limitations on the genotyping platform used and sometimes because of the biological characteristics of the species. The absence of simplex markers observed in chromosome 11 in the landrace Tanzania (Mollinari et al., 2020), probably caused by a recent nondisjunction of sister chromatids in meiosis II in one of Tanzania's parents, consists of a good example of a biological characteristic that poses a limitation to this approach. Such biological constraints highlight the need for alternative mapping strategies that can accommodate the complexities of polyploid genomes and provide a more accurate portrayal of their genetic architecture.

5.6 Parental Integration and Dosage-Based Maps

Despite the substantial theoretical progress in polyploid linkage analysis made from the 1920s to the 1960s, the construction of genetic linkage maps well into the late 1990s continued to rely heavily on simplex markers. Early efforts to integrate multi-dose or multiplex markers into mapping studies of crops like sugarcane and alfalfa highlighted the promise of these methodologies. However, these initial endeavors were hindered by the technological limitations of the time (da Silva 1993a, b; Yu and Pauls 1993; Da Silva et al. 1995; Guimarães et al. 1997). Regardless the limitations, Ripol et al. (1999) laid foundational work by dissecting the use of multiplex markers in genetic mapping for autopolyploids across arbitrary ploidy levels, although initially focusing on markers informative in just one parent. In order to expand the use of informative markers to both parents, Hackett et al. (1998) conducted a simulation study establishing formulae to compute the recombination fraction, standard error, and test power for all possible combinations of simplex and duplex markers in tetraploid species, marking a significant step toward comprehensive parental integration and dosage-based mapping.

Following these initial works, Luo et al. (2001) presented the use of dominant and codominant markers scored in an autotetraploid population for genetic map construction while Hackett et al. (2003) laid out marker ordering procedures. These efforts culminated in the development of Tetraploidmap, the first linkage mapping software developed specifically for autotetraploid species (Hackett 2003). Additionally, a series of studies aimed at modeling polyploid genetics complexities, such as double reduction, multivalent formation, and preferential pairing, were presented (Wu et al. 2001a, b; Wu and Ma 2005), providing a deeper understanding of polyploid inheritance.

The development of multilocus maps in autotetraploids represented a pivotal leap forward in polyploid genetic mapping. Luo et al. (2004) laid a comprehensive theoretical groundwork for linkage analysis in autotetraploids, which was further advanced by Leach et al. (2010) through the proposal of a definitive multilocus tetrasomic linkage analysis using Hidden Markov Models (HMM). The significance of multilocus analysis in enhancing our understanding of polyploid genetics and its broader implications will be further explored in this chapter.

The advent of high throughput technologies marked a significant evolution in genetic mapping, enabling the detailed assessment of allelic variants through SNP dosage information. This breakthrough was exemplified in the work of Hackett et al. (2013), who, leveraging the capabilities of the Infinium 8303 potato SNP array (Felcher et al. 2012), advanced the methodologies initially proposed by Luo et al. (2001). They constructed a comprehensive genetic linkage map for a biparental autotetraploid potato population encompassing 190 individuals, utilizing SNP dosage data to map a total of 3839 markers, which was used in a QTL mapping study for several important traits in potato (Hackett et al. 2014). These endeavors led to the development of an enhanced version of Tetraploidmap named TetraploidSNPMap (Hackett et al. 2017— https://www.bioss.ac.uk/knowledge-exchange/ software/TetraploidSNPMap). As the number of markers increased, there arose a need for faster and more accurate ordering algorithms. The introduction of the multidimensional scaling (MDS) algorithm for genetic mapping, along with the fitting of a principal curve to assess optimum marker order within a linkage group (Preedy and Hackett 2016), significantly propelled the advancement of genetic mapping in polyploids. This methodology facilitated the ordering and estimation of maps for thousands of markers, culminating in the creation of the R package MDSmap (https://cran.r-project.org/ package=MDSMap).

Through the application of mapping methodologies developed specifically for polyploids, studies have elucidated the genetic mechanisms underlying complex phenomena such as double reduction in potatoes (Bourke et al. 2015), preferential pairing in roses (Bourke et al. 2017), and hexasomic inheritance in chrysanthemums (van Geest et al. 2017b). The latter, utilized a biparental population consisting of 405 offspring individuals, enabling the detailed examination of genetic inheritance patterns. Using the same population, van Geest et al. (2017a) published the first integrated genetic map for an autohexaploid species. This map featured 30,312 segregating SNPs, covering 9 homology groups, and successfully identifying 107 out of the 108 expected homologs. Further refining the methodologies employed in this study, Bourke et al. (2018a) released the R package polymapR (https://cran.r-project.org/package=polymapR), which was designed to create linkage maps using dosage-based markers in outcrossing diploid, autotriploid, autotetraploid, and autohexaploid species, as well as segmental allotetraploids.

The advancements in genetic mapping for polyploid species, notably facilitated by software such as TetraploidSNPMap and polymapR, marked a significant leap forward for the polyploid research community. However, these tools initially lacked the capability for multipoint mapping estimation for high ploidy levels, specifically hexaploids, a gap that was addressed by the introduction of MAPpoly (Mollinari and Garcia 2019, https://cran.r-project.org/package=mappoly). This R package, specifically designed for constructing genetic maps in both diploids and autopolyploids across even ploidy levels up to 8 using HMMs and up to 12 through two-point simplification, represented a crucial advancement in the field. The implementation of MAPpoly facilitated the development of the first integrated multilocus genetic map in sweetpotato, which will be presented in the next section. This development not only highlighted the continuous innovation within polyploid genetics, but also set the stage for the detailed exploration and understanding of complex genetic architectures in such species.

5.7 Linkage Analysis in the Cultivated Sweetpotato

The genetic mapping in sweetpotato spans decades, evolving through various phases of scientific discovery and technological advancement. In this section, we will explore the early efforts and the progressive development of marker technologies that have shaped our understanding of genetic mapping in sweetpotato.

5.7.1 Historical Context of Genetic Mapping in Sweetpotato

In the early 1990s, Hong and Thompson (1994) conducted a preliminary analysis on four biparental crosses using RAPD markers to examine the genetic structure of parental and offspring individuals. They identified several primers that generated polymorphic bands, indicating linkage among them, and confirming their segregation according to the Mendelian ratios expected for a hexaploid organism. This early work demonstrated the feasibility of constructing a genetic linkage map for sweetpotato. Building on this foundation, Thompson et al. (1997) further analyzed 100 RAPD markers in two biparental populations, finding 74 markers that segregated at a 1:1 ratio and identifying five linked pairs of markers. This effort culminated in the first genetic linkage map for the hexaploid sweetpotato, featuring 188 RAPD markers across 18 and 16 linkage groups for each parent, respectively, with estimated total map lengths ranging from 173.1 to 265.4 cM (Ukoskit and Thompson 1997). Their method, assessing the type of polyploidy in sweetpotato through the analysis of simplex versus non-simplex markers and the interaction phases between simplex markers, supported the autohexaploid model of sweetpotato that corroborates with contemporary studies.

Subsequent work by Kriegner et al. (2003) employed AFLP markers within a biparental population to develop a genetic linkage map, utilizing the two-way pseudo-testcross strategy as described by Wu et al. (1992) and Grattapaglia and Sederoff (1994). The resulting map spanned 3655.6 cM and 3011.5 cM for the female and male parents, respectively, organized into 90 and 80 linkage groups. These were further aligned into 15 homologous linkage groups using multiplex markers. The authors analyzed the simplex versus multiplex marker ratio alongside the ratio of coupling versus repulsion linkage phases to determine the polyploidy type in sweetpotato. Their findings indicated a predominant polysomic inheritance of homologous chromosomes, once again affirming

the autohexaploid nature of sweetpotato, though they noted a minor degree of preferential chromosome pairing. Additionally, the linkage map for the female parent was later detailed and utilized in research mapping resistance to virus diseases in sweetpotato (Mwanga et al. 2002), showcasing the map's practical application in addressing specific agricultural challenges.

Using another mapping population composed of 240 individuals of a cross between the cream-fleshed African landrace 'Tanzania' and the US orange-fleshed cultivar 'Beauregard', Cervantes-Flores et al. (2008) made a significant contribution by constructing a detailed genetic linkage map using 3695 AFLP, which spanned 5792 cM and 5276 cM for female and male genomes, respectively, across 86 and 90 linkage groups. Their findings pointed to the presence of distorted segregation among markers of different dosages, suggesting some level of preferential pairing. However, due to the observed segregation ratios and the proportion of simplex to multiplex markers, they concluded that strict allopolyploidy could be ruled out, suggesting a complex inheritance pattern more aligned with an autohexaploid organism. In parallel, other studies also contributed to the genetic mapping of sweetpotato. Chang et al. (2009) utilized 37 and 47 inter simple sequence repeat (ISSR) markers to represent 479.8 and 853.5 cM of each parent genome, and Li et al. (2010) utilized 801 sequence-related amplified polymorphism (SRAP) markers that showed polymorphisms in the 240 individuals of the progeny to represent 81 and 66 linkage groups with total lengths of 5802.46 and 3967.90 cM for each parent, respectively.

Zhao et al. (2013) published what was then the densest genetic linkage map of the sweetpotato genome, utilizing 4031 AFLP and SSR markers. These markers were distributed across 90 linkage groups for both parents, covering lengths of 8184.5 and 8151.7 cM, respectively. This achievement marked the first time a genetic linkage map covered all linkage groups for both parents in sweepotato, representing a significant advancement for genetic research in this crop.

Two years later, Monden et al. (2015) developed a genetic linkage map using 98 progeny individuals and 246 retrotransposon insertion polymorphism (RIP) markers. They successfully reconstructed 43 and 47 linkage groups for each parent, covering 931.5 cM and 734.3 cM of the parental genomes, respectively. This work demonstrated the usefulness and efficiency of retrotransposon-based molecular markers in constructing a genetic map for a polyploid species, highlighting the evolving landscape of genomic tools in sweetpotato research.

The first genetic linkage map that utilized data from a high throughput genotyping platform in sweetpotato was published by Shirasawa et al. (2017). This linkage map featured 28,087 markers obtained with the double-digest RAD-Seq (ddRAD-Seq) technology that spanned all 90 sweetpotato linkage groups, along with an additional six groups. These were subsequently consolidated into 15 homology groups, covering 33,020.4 cM of the sweetpotato genome. Notably, this map was also the first for sweetpotato to be constructed with the aid of a reference genome, utilizing the wild diploid relative *Ipomoea trifida* as a basis for anchoring and calling variants. It was also the first sweetpotato linkage map developed jointly for both parents. It remains the only map based on a population derived from a self-fertilizing individual, breaking a biological barrier and setting a precedent for future genetic research in sweetpotato.

In (2020), Mollinari et al. published the most comprehensive genetic linkage map for sweetpotato to date, using a biparental population of 315 individuals from a cross between 'Beauregard' and 'Tanzania' (BT), the same parents studied by Cervantes-Flores et al. (2008), but in a reciprocal cross. This ultra-dense map featured 30,684 SNP markers distributed across 15 linkage groups, spanning 2708.36 cM of the sweetpotato genome. It was the first genetic linkage map constructed for both parents simultaneously, accounting for the hexaploid nature and multiple dosages of sweetpotato through a refined analytical approach, detailed

by Mollinari and Garcia (2019). The process started with pairs of markers and progressively incorporating them into an HMM framework to re-estimate recombination fractions and linkage phases. The study revealed that homolog pairing predominantly occurs randomly, supporting sweetpotato's classification as an autohexaploid. Analysis of meiotic configurations revealed that most gametes formed through pairing and recombination between two homologs, with a smaller percentage showing configurations that involved three to six homologs, indicative of multivalent formations. This pattern was consistent across both 'Beauregard' and 'Tanzania' parents, with a noted correlation between multivalent formations and the length of linkage groups. An interactive version of the BT map can be accessed at https://gt4sp-genetic-map. shinyapps.io/bt_map/. Figure 5.4 features a screenshot from a web-based application that illustrates the meiotic processes implicated in forming homology group 1 for individual BT05.221. The application allows the exploration of any homology group across different offspring individuals, facilitating a comprehensive understanding of their gamete formation.

Oloka et al. (2021) reported another sweetpotato map based on the same cross used by Cervantes-Flores et al. (2008) ('Tanzania' × 'Beauregard') but based on a high-throughput GBS genotyping platform and using a similar methodology employed by Mollinari et al. (2020). The authors assembled a dense, high-quality genetic linkage map that comprised 14,813 markers, covering 2120.5 cM of the sweetpotato genome, and representing all 15 linkage groups. This linkage map was slightly shorter than the one reported by Mollinari et al. (2020), with a smaller progeny size and approximately half the number of markers in the final linkage map. As marker technologies have become more affordable and accessible, there has been a significant surge in studies incorporating genomic information in sweetpotato research. Kim et al. (2017) and Li et al. (2018) developed maps using EST-SSR and

Haplotype estimation and multivalent evidence

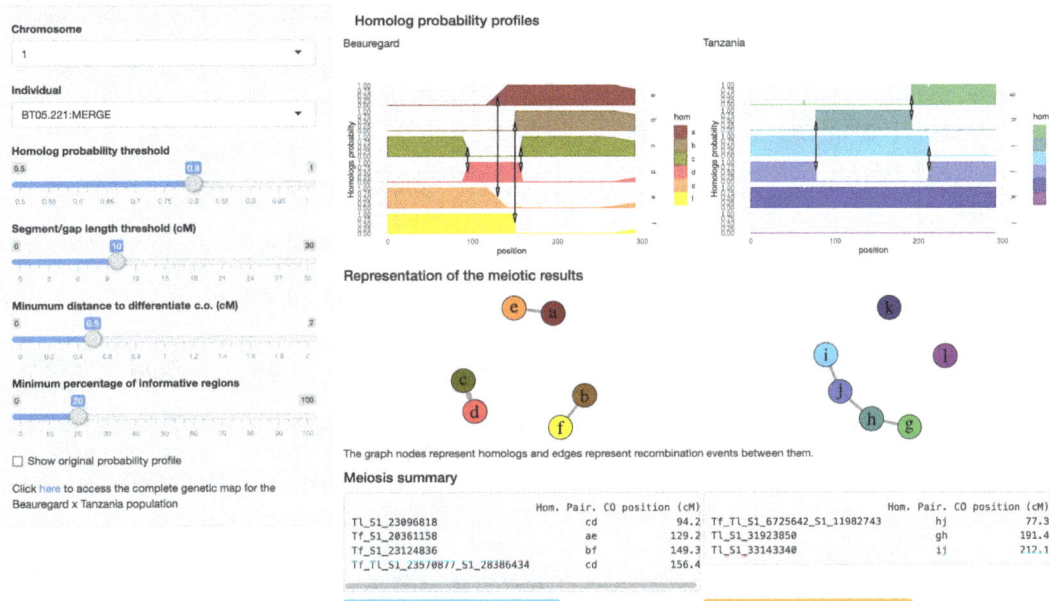

Fig. 5.4 Interactive display of haplotype estimation and multivalent evidence in gamete formation analysis of homology group 1 for individual BT05.221. The color-coded probability profiles in the figure showcase homologs a-f for 'Beauregard' and g-l for 'Tanzania'. Arrows within these profiles mark crossing-over points, with their positions presented in the accompanying table. The network diagram employs the same colors to illustrate recombination chains; for instance, homologs c and d are connected by two edges, indicating a double crossover, whereas homologs i, j, h, and g form a recombination chain, revealing multivalent pairing evidence. The left panel provides user-controlled parameters for detecting these genetic events, with the methodology and analytical detail explained in Mollinari et al. (2020). This tool and its capabilities for exploring genetic linkages in BT offspring can be accessed at https://gt4sp-genetic-map.shinyapps.io/offspring_haplotype_BT_population/

SRAP markers, respectively. Sasai et al. (2019) used retrotransposons, SSR, and SNP markers for their linkage map reconstruction. Ma et al. (2020) constructed the first genetic linkage map for purple sweetpotato using SSR markers, followed by Haque et al. (2020), who employed ddRAD-Seq markers. Meng et al. (2021) created the densest SSR-based linkage map to date for sweetpotato, while Yan et al. (2022) developed the second map for purple sweetpotato, this time utilizing SNP-based markers. Most recently, Zheng et al. (2023) achieved the longest coverage of the sweetpotato genome, with 10,146 SSR markers covering over 18,000 cM. A comprehensive list of all published sweetpotato genetic linkage maps is available in Table 5.1.

Several key factors have contributed to the increased density and quality of these recent genetic linkage maps for sweetpotato. These include the reduced cost of genotyping technologies, a surge in multi-omics research efforts, and the availability of tools and resources that are appropriate for polyploid genotyping and data analysis. Such resources include sequencing protocols tailored for polyploids (Wadl et al. 2018; Mollinari et al. 2020), high-quality reference genomes (Wu et al. 2018), and advanced statistical genetics methods for conducting accurate linkage analysis in polyploids (Bourke et al. 2018a, b; Mollinari and Garcia 2019; Zheng et al. 2021).

Table 5.1 Sweetpotato genetic maps published to date

Publications	Marker type	P1	P2	Progeny size	Map length P1[a]	Map length P2[a]	Markers
Ukoskit and Thompson (1997)	RAPD	Vardaman	Regal	76	265.4	173.1	188
Kriegner et al. (2003)	AFLP	Tanzania	Bikilamaliya	94	3655.6	3011.5	1276
Cervantes-Flores et al. (2008)	AFLP	Tanzania	Beauregard	240	5792	5276	3695
Li et al. (2010)	SRAP	Luoxushu 8	Zhengshu 20	240	5802.46	3967.9	801
Zhao et al. (2013)	AFLP/SSR	Xushu 18	Xu 781	202	8184.5	8151.7	4031
Monden et al. (2015)	RIP[b]	PSL[c]	90IDN-47	98	931.5	734.3	246
Shirasawa et al. (2017)	ddRAD-Seq (SNP)	Xushu 18[d]	Xushu 18[d]	142	33,020.4[e]	33,020.4[e]	28,087
Kim et al. (2017)	EST-SSR	Yeseumi	Annobeny	137	1508.1	1508.1	210
Li et al. (2018)	SRAP	Luoxushu 8	Zhengshu 20	240	10,188	9165	2504
Mollinari et al. (2020)	GBSpoly (SNP)	Beauregard	Tanzania	298	2708.3[f]	2708.3[f]	30,684
Oloka et al. (2021)	GBSpoly (SNP)	Tanzania	Beauregard	244	2120[f]	2120[f]	14,813
Haque et al. (2020)	ddRAD-Seq (SNP)	Konaishin	Akemurasaki	94	4726	4726	15,747
Sasai et al. (2019)	RT/SSR/SNP	J-Red	Choshu	113	13,247	12,242	12,349
Ma et al. (2020)	SSR	Jizishu 1	Longshu 9	300	3974.24	5163.35	1057
Meng et al. (2021)	SSR	Luoxushu 8	Zhengshu 20	240	13,299.9	11,122.9	8066
Yan et al. (2022)	SLAF-seq (SNP)	Xuzishu 8	Meiguohong	274	2233.66	2233.66	3178
Zheng et al. (2023)	SSR	Xushu 18	Xu 781	50	18,263.5	18,043.7	10,146

[a]The map length refers to the haploid map, except for (Mollinari et al. 2020) and (Oloka et al. 2021)
[b]Retrotransposon Insertion Polymorphisms
[c]Purple Sweet Lord
[d]Self-pollination of Xushu 18
[e]Map length refers to the integrated linkage map
[f]Map length refers to the joint linkage map

5.7.2 Advanced Genetic Mapping Techniques for Sweetpotato

5.7.2.1 Multilocus Analysis: Importance and Methods

As presented so far, linkage analysis has been used to study patterns of inheritance, meiotic landscapes (Bourke et al. 2015), and to provide a foundational framework for subsequent genetic and genomic analyses, including QTL mapping (Doerge 2002), evolutionary studies (Ahn and Tanksley 1993; Huang and Rieseberg 2020), and the assembly of genomes (Lewin et al. 2009).

A significant challenge encountered in genetic mapping is the issue of missing data, which not only pertains to data failing to meet quality control threshold filters, but also includes data that inherently offers only partial genotype information of the loci under study. This challenge was notably prevalent in diploid mapping populations when using dominant markers, such as AFLPs and RAPD. For instance, in an F_2 cross, the anticipated 1:2:1 segregation ratio for AA:Aa:aa genotypes is effectively reduced to a 3:1 ratio due to these markers' inability to distinguish between AA and Aa genotypes. Consequently, identifying recombination events with such markers necessitates resolving these combined classes through numerical algorithms like Expectation–Maximization (EM) (Dempster et al. 1977). This phenomenon, a well-documented property in the study of natural systems within statistics for decades, was explored in its application to linkage analysis in the influential work by Mather (1957). The impact of missing data becomes especially apparent in the context of linkage analysis between two positions in the genome. In such instances, the ability to detect recombination events is constrained by the information available from the experimental population concerning only those two specific loci. Typically referred to as two-point analysis, this approach involves examining all marker pairs across the genome to estimate the recombination fraction or genetic distance between them (Liu 1998).

Another frequent issue in linkage analysis is the significant amount of noise observed in the dataset, which can propagate to downstream analyses. This issue may arise from duplications during the polymerase chain reaction (PCR) that lead to biases and sequencing errors, which culminates in inflated and incorrect linkage maps and has become more prevalent with the advent of high-throughput sequencing technologies (Taniguti et al. 2022). While these technologies have revolutionized genetic research by providing thousands of markers at a reasonable cost, they sometimes also introduce a lower signal-to-noise ratio. This decrease in data clarity can complicate analyses, making it challenging to distinguish between meaningful genetic signals and mere background noise. The problem is especially pronounced in polyploids, where many genotypes are possible and the genome's complexity adds layers of difficulty to the precise interpretation of data (Gemenet et al. 2020; Liao et al. 2021).

Addressing the challenges posed by noise and missing data in genetic datasets demands the adoption of a joint analysis of groups of genomic loci, a process that is fundamental to enhancing the clarity and utility of genetic information. This method, known as multilocus or multipoint analysis, capitalizes on the simultaneous use of information from multiple markers, facilitating information sharing among markers situated closely on the genome. This strategy has been a cornerstone of genetic linkage analysis since the early 1980s, marked by groundbreaking contributions from researchers like Thompson (1984), Lathrop and Lalouel (1984), and Lathrop et al. (1985), along with the seminal work by Lander and Green (1987). They introduced Hidden Markov Models for reconstructing genetic maps, offering a robust framework for addressing missing data issues, as thoroughly analyzed by Jiang and Zeng (1997) in various diploid experimental populations.

Multilocus methods have become increasingly important in polyploids. As illustrated in Fig. 5.3, the number of genotypes that can be generated from biparental crosses is staggering. A fully informative marker, i.e., a marker capable of distinguishing all possible genotypic classes, is practically nonexistent (Leach et al. 2010). Despite the widespread use of codominant markers such as SNPs, several genotypic classes are still collapsed into one class (Fig. 5.3). Therefore, applying multilocus analysis in polyploids becomes a crucial tool for overcoming the limitations of the less informative nature of commonly used markers. Furthermore, it is well-documented that as the ploidy level increases, differentiating between genuine genetic signals and background noise becomes more challenging (Liao et al. 2021), highlighting the critical need for advanced analysis techniques like multilocus analysis.

In the realm of autotetraploid species, specifically designed and successfully applied multilocus methods have facilitated the construction of genetic maps, advancing our understanding of these organisms (Hackett 2001, 2003; Hackett et al. 2013; Leach et al. 2010; Xie and Xu 2000; Zheng et al. 2016, 2021). These studies not only highlight the utility of multilocus approaches in mapping autotetraploid species but also explore the complexities of tetrasomic inheritance, including the phenomenon of double reduction. To date, the only multilocus method available for constructing genetic maps in hexaploids was introduced by Mollinari and Garcia (2019). This method can theoretically handle a biparental cross between parents of any even ploidy level. In their work, the authors derived general equations to calculate recombination fractions between pairs of markers, considering all possible phase configurations. This two-point-based method was integrated with a general multipoint HMM approach and combined with a sequential algorithm to narrow down the search space for phase configurations. The R package MAPpoly incorporates the multipoint algorithm along with a suite of additional tools, offering a cutting-edge linkage analysis system specifically designed for outcrossing species with even ploidy levels varying from 2 to 8, when using the multilocus approach, and up to 12 when using two-point based algorithms.

A crucial benefit of the multilocus approach lies in its ability to utilize information propagated along the chromosome chain to refine loci genotypes, thereby rectifying potential dosage misclassifications and genotype inaccuracies. This refinement process, described in detail by Mollinari and Garcia (2019), leverages the analytical power of HMMs to update the posterior probabilities of genotypes. HMMs iteratively adjust these probabilities based on observed data and the sequence of events leading to them, thus effectively correcting genotype assignments. This correction mechanism was utilized in the genetic mapping efforts of Mollinari et al. (2020) and Oloka et al. (2021) and has been implemented in the genetic mapping software MAPpoly. Through this methodology, HMMs provide a robust framework for enhancing the accuracy of genetic maps, ensuring the integrity of the data analysis process.

5.7.2.2 High Throughput Data and Its Impacts in Sweetpotato Mapping

The advent of high-throughput sequencing technologies has fundamentally transformed genetic mapping in polyploid species, offering the capability to generate an unprecedented volume of molecular marker data. This surge in data availability has opened new avenues for genetic research, allowing for the detailed exploration of complex genomes that were once considered too challenging to analyze effectively.

High-throughput technologies have facilitated the identification of a broad spectrum of genetic variations, providing a rich dataset for constructing more accurate and comprehensive genetic maps. The densest and highest quality linkage maps available for sweetpotato to date were constructed using molecular markers that were obtained using high throughput technologies (Shirasawa et al. 2017; Mollinari et al. 2020; Oloka et al. 2021; Haque et al. 2020; Sasai et al. 2019). Some of these maps, as well as other studies in sweetpotato, could benefit from the information contained in one or multiple reference genomes (Wu et al. 2018), which could only be assembled by using high throughput sequencing technologies. Such developments are crucial for advancing our understanding of the sweetpotato genome.

The abundance of molecular markers provided by these new technologies makes it feasible to rigorously filter out markers that exhibit low signal-to-noise ratios or are plagued by specific issues encountered in GBS, such as allele dropout, PCR amplification biases, or sequencing errors. However, it is crucial to avoid discarding markers that deviate from expected patterns, such as Mendelian segregation, without thorough investigation (Mollinari et al. 2020). Such deviations may not be artifacts but rather indicators of underlying biological phenomena. Identifying whether unexpected patterns are confined to specific chromosome segments or individuals within the population can provide valuable insights. The power of high-throughput sequencing is undeniable, but it also demands substantial computational resources and sophisticated algorithms to navigate the complexities inherent in polyploid analysis, underscoring the need for continuous advancements in bioinformatics to fully leverage the potential of this technology in genetic mapping (Taniguti et al. 2022).

5.8   Final Remarks

Although the concept of genetic maps dates to the origins of genetics, they remain vital tools for elucidating the genome behavior during meiosis. This is particularly true in polyploids, where meiosis presents added complexity. Studies on potato by Bourke et al. (2015) and Pereira et al. (2021), blueberry by Cappai et al. (2020), and sweetpotato by Mollinari et al. (2020) exemplify the utility of genetic maps in unraveling the meiotic processes and its characteristics in polyploid organisms. Furthermore, applying HMM-based multilocus analysis not only improves the construction of these maps, but also offers a mean to correcting inaccuracies or misclassifications within datasets, a common issue in organisms with high ploidy levels, such as sweetpotato. As highlighted by Mollinari et al. (2020), a genetic map transcends the mere linear arrangement of markers along linkage groups; it elucidates the inheritance patterns governing the genome transfer from parents to offspring, with the mapping method's capacity to estimate haplotypes across generations and providing a comprehensive characterization of this transmission process.

References

Ahn S, Tanksley SD (1993) Comparative linkage maps of the rice and maize genomes. Proc Natl Acad Sci 90:7980–7984. https://doi.org/10.1073/pnas.90.17.7980

Ai-xian L, Qing-chang L, Qing-mei W et al (2010) Establishment of molecular linkage maps using SRAP markers in sweet potato. Acta Agron Sin 36:1286–1295. https://doi.org/10.3724/SP.J.1006.2010.01286

Amadeu RR, Muñoz PR, Zheng C, Endelman JB (2021) QTL mapping in outbred tetraploid (and diploid) diallel populations. Genetics 219:iyab124. https://doi.org/10.1093/genetics/iyab124

Balsalobre TWA, Pereira GS, Margarido GRA et al (2017) GBS-based single dosage markers for linkage and QTL mapping allow gene mining for yield-related traits in sugarcane. BMC Genomics 18:1–19. https://doi.org/10.1186/s12864-016-3383-x

Bink M, Jansen J, Madduri M et al (2014) Bayesian QTL analyses using pedigreed families of an outcrossing species, with application to fruit firmness in apple. Theor Appl Genet 127:1073–1090. https://doi.org/10.1007/s00122-014-2281-3

Bourke PM, Arens P, Voorrips RE, Esselink GD, Koning-Boucoiran CFS, van't Westende WPC, Santos Leonardo T, Wissink P, Zheng C, van Geest G, Visser RGF, Krens, Frans A, Smulders MJM, Maliepaard C (2017) Partial preferential chromosome pairing is genotype dependent in tetraploid rose. Plant J 90(2):330–343. https://doi.org/10.1111/tpj.13496

Bourke PM, Voorrips RE, Visser RGF, Maliepaard C (2015) The double-reduction landscape in tetraploid potato as revealed by a high-density linkage map. Genetics 201:853–863. https://doi.org/10.1534/genetics.115.181008

Bourke PM, Van Geest G, Voorrips RE et al (2018a) PolymapR—linkage analysis and genetic map construction from F1 populations of outcrossing polyploids. Bioinformatics 34:3496–3502. https://doi.org/10.1093/bioinformatics/bty371

Bourke PM, Voorrips RE, Visser RGF, Maliepaard C (2018b) Tools for genetic studies in experimental populations of polyploids. Front Plant Sci 9:1–17. https://doi.org/10.3389/fpls.2018.00513

Bourke PM, Voorrips RE, Hackett CA et al (2021) Detecting quantitative trait loci and exploring chromosomal pairing in autopolyploids using polyqtlR. Bioinformatics 37:3822–3829. https://doi.org/10.1093/bioinformatics/btab574

Broman KW, Sen Ś (2009) A guide to QTL mapping with R/qtl. Springer. https://doi.org/10.1007/978-0-387-92125-9

Buckler ES, Holland JB, Bradbury PJ et al (2009) The genetic architecture of maize flowering time. Science 325:714–718. https://doi.org/10.1126/science.1174276

Cai G, Yang Q, Yi B et al (2014) A complex recombination pattern in the genome of allotetraploid brassica napus as revealed by a high-density genetic map. PLoS ONE 9:e109910. https://doi.org/10.1371/journal.pone.0109910

Cappai F, Amadeu RR, Benevenuto J et al (2020) High-resolution linkage map and QTL analyses of fruit firmness in autotetraploid blueberry. Front Plant Sci 11:1–11. https://doi.org/10.3389/fpls.2020.562171

Cervantes-Flores JC, Yencho GC, Kriegner A et al (2008) Development of a genetic linkage map and identification of homologous linkage groups in sweetpotato using multiple-dose AFLP markers. Mol Breeding 21:511–532. https://doi.org/10.1007/s11032-007-9150-6

Chakraborty N, Bae J, Warnke S et al (2005) Linkage map construction in allotetraploid creeping bentgrass (Agrostis stolonifera L.). Theor Appl Genet 111:795–803. https://doi.org/10.1007/s00122-005-2065-x

Chang KY, Lo HF, Lai YC et al (2009) Identification of quantitative trait loci associated with yield-related traits in sweet potato (Ipomoea batatas). Bot Stud 50:43–55

Clark L V, Lipka AE, Sacks EJ (2019) polyRAD: genotype calling with uncertainty from sequencing data in polyploids and diploids. G3 Genes|Genomes|Genetics 9:g3.200913.2018. https://doi.org/10.1534/g3.118.200913

Pereira GS, Gemenet DC, Mollinari M et al (2020) Multiple QTL mapping in autopolyploids: a random-effect model approach with application in a hexaploid sweetpotato full-sib population. Genetics 215:579–595. https://doi.org/10.1534/genetics.120.303080

da Pereira GS, Mollinari M, Schumann MJ et al (2021) The recombination landscape and multiple QTL mapping in a Solanum tuberosum cv. 'Atlantic'-derived F1 population. Heredity (Edinb). https://doi.org/10.1038/s41437-021-00416-x

da Silva JAG (1993a) A methodology for genome mapping of autopolyploids and its application to sugarcane (Saccharum spp.). Cornell University

Da Silva JGA (1993b) A methodology for genome mapping of autopolyploids and its application to sugarcane {\it Saccharum} spp. Cornell University

Da Silva JAG, Honeycutt RJ, Burnquist W et al (1995) Saccharum spontaneum L. "SES 208" genetic linkage map combining RFLP- and PCR-based markers. Mol Breed 1:165–179

Dempster AP, Laird NM, Rubin DB (1977) Maximum likelihood from incomplete data via the EM algorithm. J Roy Stat Soc: Ser B (methodol) 39:1–22

Deo TG, Ferreira RCU, Lara LAC et al (2020) High-resolution linkage map with allele dosage allows the identification of regions governing complex traits and apospory in guinea grass (Megathyrsus maximus). Front Plant Sci 11:15. https://doi.org/10.3389/fpls.2020.00015

Doerge RW (2002) Mapping and analysis of quantitative trait loci in experimental populations. Nat Rev Genet 3:43–52. https://doi.org/10.1038/nrg703

Doerge RW, Zeng Z-B, Weir BS (1997) Statistical issues in the search for genes affecting quantitative traits in experimental populations. Stat Sci 12:195–219

Elandt-Johnson RC (1967) Equilibrium conditions in polysomic inheritance for a panmictic population. Bull Math Biophys 29:437–449. https://doi.org/10.1007/BF02476583

Felcher KJ, Coombs JJ, Massa AN et al (2012) Integration of two diploid potato linkage maps with the potato genome sequence. PLoS ONE 7:1–11. https://doi.org/10.1371/journal.pone.0036347

Ferreira RCU, Lara LAC, Chiari L et al (2019) Genetic mapping with allele dosage information in tetraploid urochloa decumbens (Stapf) R. D. webster reveals insights into spittlebug (Notozulia Entreriana Berg) resistance. Front Plant Sci 10:92. https://doi.org/10.3389/fpls.2019.00092

Fisher RA (1943) Allowance for double reduction in the calculation of genotype frequencies with polysomic inheritance. Ann of Eugen 12:169–171. https://doi.org/10.1111/j.1469-1809.1943.tb02320.x

Fisher RA (1947) The theory of linkage in polysomic inheritance. Philos Trans R Soc Lond 233:55–87

Fisher RA, Mather K (1940) Non-lethality of the mid factor in Lythrum Salicaria. Nature 146:521

Gallais A (2003) Quantitative genetics and breeding methods in autopolyploid plants. Institut national de la recherche agronomique, Paris

van Geest G, Bourke PM, Voorrips RE, Marasek-Ciolakowska A, Liao Y, Post A, van Meeteren U, Visser RGF, Maliepaard C, Arens P (2017a) An ultra-dense integrated linkage map for hexaploid chrysanthemum enables multi-allelic QTL analysis. Theor Appl Genet 130(12):2527–2541. https://doi.org/10.1007/s00122-017-2974-5

van Geest G, Voorrips RE, Esselink D, Post A, Visser RGF, Arens P (2017b) Conclusive evidence for hexasomic inheritance in chrysanthemum based on analysis of a 183 k SNP array. BMC Genomics 18(1):1–12. https://doi.org/10.1186/s12864-017-4003-0

Gemenet DC, Lindqvist-Kreuze H, De Boeck B et al (2020) Sequencing depth and genotype quality: accuracy and breeding operation considerations for genomic selection applications in autopolyploid crops. Theor Appl Genet 133:3345–3363. https://doi.org/10.1007/s00122-020-03673-2

Gerard D, Ferrão LFV, Garcia AAF, Stephens M (2018) Genotyping polyploids from messy sequencing data. Genetics 210:789–807. https://doi.org/10.1534/genetics.118.301468

Grattapaglia D, Sederoff R (1994) Genetic linkage maps of eucalyptus grandis and eucalyptus Urophylla using a pseudo-testcross: mapping strategy and Rapd markers. Genetics 137:1121–1137. https://doi.org/10.1093/genetics/137.4.1121

Guimarães CT, Sills GR, Sobral BWS (1997) Comparative mapping of Andropogoneae: Saccharum L. (sugarcane) and its relation to sorghum and maize. Proc Natl Acad Sci USA 94:14261–14266

Hackett CA (2003) Tetraploidmap: construction of a linkage map in autotetraploid species. J Hered 94:358–359. https://doi.org/10.1093/jhered/esg066

Hackett CA, Bradshaw JE, McNicol JW (2001) Interval mapping of quantitative trait loci in autotetraploid species. Genetics 159:1819–1832. https://doi.org/10.1093/genetics/159.4.1819

Hackett CA, Pande B, Bryan GJ (2003) Constructing linkage maps in autotetraploid species using simulated annealing. Theor Appl Genet 106:1107–1115. https://doi.org/10.1007/s00122-002-1164-1

Hackett CA, Bradshaw JE, Bryan GJ (2014) QTL Mapping in autotetraploids using Snp dosage information. Theor Appl Genet 127:1885–1904. https://doi.org/10.1007/s00122-014-2347-2

Hackett CA, Boskamp B, Vogogias A et al (2017) TetraploidSNPMap: software for linkage analysis and QTL mapping in autotetraploid populations using SNP dosage data. J Hered 108:438–442. https://doi.org/10.1093/jhered/esx022

Hackett CA, Bradshaw JE, Meyer RC et al (1998) Linkage analysis in tetraploid species: a simulation study. Genet Res 71:143–153. https://doi.org/10.1017/S0016672398003188

Hackett CA, McLean K, Bryan GJ (2013) Linkage analysis and Qtl mapping using Snp dosage data in a tetraploid potato mapping population. PLoS One 8:e63939. https://doi.org/10.1371/journal.pone.0063939

Hackett CA (2001) A comment on Xie and Xu: 'Mapping quantitative trait loci in tetraploid species.'

Genet Res 78:187–189. https://doi.org/10.1017/S0016672301005262

Haldane J (1919) The combination of linkage values, and the calculation of distance between linked factors. J Genet 8:299–309

Haldane JBS (1930) Theoretical genetics of autopolyploids. J Genet 22:359–372. https://doi.org/10.1007/BF02984197

Haque E, Tabuchi H, Monden Y, Suematsu K, Shirasawa K et al (2020) Qtl analysis and Gwas of agronomic traits in sweetpotato (Ipomoea Batatas L.) using genome wide Snps. Breed Sci 70:283–291. https://doi.org/10.1270/jsbbs.19099

Hong LL, Thompson PG (1994) Genetic linkages of rapd markers in sweetpotato. HortScience 29:727d–7727. https://doi.org/10.21273/hortsci.29.7.727d

Huang K, Rieseberg LH (2020) Frequency, origins, and evolutionary role of chromosomal inversions in plants. Front Plant Sci 11:296. https://doi.org/10.3389/fpls.2020.00296

Huang BE, Verbyla KL, Verbyla AP et al (2015) MAGIC populations in crops: current status and future prospects. Theor Appl Genet 128:999–1017

Jiang C, Zeng Z-B (1997) Mapping quantitative trait loci with dominant and missing markers in various crosses from two inbred lines. Genetica 101:47–58. https://doi.org/10.1023/a:1018394410659

Kim J-H, Chung IK, Kim K-M (2017) Construction of a genetic map using Est-Ssr markers and Qtl analysis of major agronomic characters in hexaploid sweet potato (Ipomoea batatas (L.) Lam. PLoS One 12:e0185073. https://doi.org/10.1371/journal.pone.0185073

Kosambi DD (1943) The estimation of map distances from recombination values. Ann Eugen 12:172–175. https://doi.org/10.1111/j.1469-1809.1943.tb02321.x

Kriegner A, Cervantes JC, Burg K et al (2003) A genetic linkage map of sweet potato [Ipomoea batatas (L.) Lam.] based on AFLP markers. Mol Breeding 11:169–185. https://doi.org/10.1023/A:1022870917230

Lander ES, Green P (1987) Construction of multilocus genetic linkage maps in humans. Proc Natl Acad Sci 84:2363–2367. https://doi.org/10.1073/pnas.84.8.2363

Lathrop GM, Lalouel JM (1984) Strategies for multilocus linkage analysis in humans. Proc Natl Acad Sci USA 81:3443–3446

Lathrop GM, Lalouel JM, Julier C, Ott J (1985) Multilocus linkage analysis in humans: detection of linkage and estimation of recombination. Am J Hum Genet 37:482–498

Lawrence JC (1929) The genetics and cytology of dahlia species. J Genet 21:125–159

Leach LJ, Wang L, Kearsey MJ, Luo Z (2010) Multilocus tetrasomic linkage analysis using hidden Markov chain model. Proc Natl Acad Sci U S A 107:4270–4274. https://doi.org/10.1073/pnas.0908477107

Lewin HA, Larkin DM, Pontius J, O'Brien SJ (2009) Every genome sequence needs a good map. Genome Res 19:1925–1928

Li A-X, Liu Q-C, Wang Q-M et al (2010) Establishment of molecular linkage maps using Srap markers in sweet potato. Acta Agron Sin 36:1286–1295. https://doi.org/10.3724/sp.j.1006.2010.01286

Li A, Qin Z, Hou F, Dong S, Wang Q (2018) Development of molecular linkage maps in sweet potato (Ipomoea Batatas L.) using sequence-related amplified polymorphism markers. Plant Breeding 137:644–654. https://doi.org/10.1111/pbr.12599

Liao Y, Voorrips RE, Bourke PM et al (2021) Using probabilistic genotypes in linkage analysis of polyploids. Theor Appl Genet 134:2443–2457. https://doi.org/10.1007/s00122-021-03834-x

Lincon SE, Daly MJ, Lander ES (1992) General information on MAPMAKER version 3.0. 54

Liu B (1998) Statistical genomics: linkage, mapping, and QTL analysis. CRC Press, Boca Raton, Florida

Luo ZW, Hackett CA, Bradshaw JE et al (2001) Construction of a genetic linkage map in tetraploid species using molecular markers. Genetics 157:1369–1385

Luo ZW, Zhang RM, Kearsey MJ (2004) Theoretical basis for genetic linkage analysis in autotetraploid species. Proc Natl Acad Sci 101:7040–7045

Lynch M, Walsh B (1998) Genetics and analysis of quantitative traits, 1st edn. Sinauer Associates

Ma Z, Gao W, Liu L, et al (2020) Identification of QTL for resistance to root rot in sweetpotato (Ipomoea batatas (L.) Lam with SSR linkage maps. BMC Genom 21:366. https://doi.org/10.1186/s12864-020-06775-9

Margarido GRA, Souza AP, Garcia AAF (2007) OneMap: software for genetic mapping in outcrossing species. Hereditas 144:78–79. https://doi.org/10.1111/j.2007.0018-0661.02000.x

Margarido GRA, Pastina MM, Souza AP, Garcia AAF (2015) Multi-trait multi-environment quantitative trait loci mapping for a sugarcane commercial cross provides insights on the inheritance of important traits. Mol Breeding 35:175. https://doi.org/10.1007/s11032-015-0366-6

Mather K (1935) Reductional and equational separation of the chromosomes in bivalents and multivalents. J Genet 30:53–78. https://doi.org/10.1007/BF02982205

Mather K (1936) Segregation and linkage in autotetraploids. J Genet 32:287–314. https://doi.org/10.1007/BF02982683

Mather K (1957) The mesurement of linkage in heredity. Methuen & Co, London

Meng Y, Zheng C, Li H et al (2021) Development of a high-density Ssr genetic linkage map in sweet potato. Crop J 9:1367–1374. https://doi.org/10.1016/j.cj.2021.01.003

Ming R (2001) Qtl analysis in a complex autopolyploid: genetic control of sugar content in sugarcane.

Genome Res 11:2075–2084. https://doi.org/10.1101/gr.198801

Mollinari M, Margarido GRA, Vencovsky R, Garcia AAF (2009) Evaluation of algorithms used to order markers on genetic maps. Heredity (edinb) 103:494–502. https://doi.org/10.1038/hdy.2009.96

Mollinari M, Garcia AAF (2019) Linkage analysis and haplotype phasing in experimental autopolyploid populations with high ploidy level using hidden markov models. G3 Genes|Genomes|Genetics 9:3297–3314. https://doi.org/10.1534/g3.119.400378

Mollinari M, Olukolu BA, Pereira G da S et al (2020) Unraveling the hexaploid sweetpotato inheritance using ultra-dense multilocus mapping. G3 Genes|Genomes|Genetics 10:281–292. https://doi.org/10.1534/g3.119.400620

Monden Y, Tahara M (2017) Genetic linkage analysis using DNA markers in sweetpotato. Breed Sci 51:41–51. https://doi.org/10.1270/jsbbs.16142

Monden Y, Hara T, Okada Y et al (2015) Construction of a linkage map based on retrotransposon insertion polymorphisms in sweetpotato via high-throughput sequencing. Breed Sci 65:145–153. https://doi.org/10.1270/jsbbs.65.145

Morgan TH (1917) The theory of the gene. Am Nat 51:513–544

Morgan TH (1911) Random segregation versus coupling in mendelian inheritance. Sci 34:384–384. https://www.science.org/doi/10.1126/science.34.873.384

Muller HJ (1914) A new mode of segregation in gregory's tetraploid primulas. Am Nat 48:508–512. https://doi.org/10.1086/521238

Mwanga ROM, Kriegner A, Cervantes-Flores JC et al (2002) Resistance to sweetpotato chlorotic stunt virus and sweetpotato feathery mottle virus is mediated by two separate recessive genes in sweetpotato. J Am Soc Hortic Sci 127:798–806. https://doi.org/10.21273/jashs.127.5.798

Nice LM, Steffenson BJ, Blake TK et al (2017) Mapping agronomic traits in a wild barley advanced backcross–nested association mapping population. Crop Sci 57:1199–1210

Oloka BM, da Silva Pereira G, Amankwaah V, Mollinari M et al (2021) Discovery of a major Qtl for rootknot nematode (meloidogyne incognita) resistance in cultivated sweetpotato (Ipomoea batatas). Theoret Appl Genet 134:1945–1955. https://doi.org/10.1007/s00122-021-03797-z

Osborn TC, Pires JC, Birchler JA et al (2003) Understanding mechanisms of novel gene expression in polyploids. Trends Genet 19:141–147. https://doi.org/10.1016/s0168-9525(03)00015-5

Pastina MM, Malosetti M, Gazaffi R et al (2011) A mixed model Qtl analysis for sugarcane multiple-harvest-location trial data. Theor Appl Genet 124:835–849. https://doi.org/10.1007/s00122-011-1748-8

Preedy KF, Hackett CA (2016) A rapid marker ordering approach for high-density genetic linkage maps in experimental autotetraploid populations using multidimensional scaling. Theor Appl Genet 129(11):2117–2132. https://doi.org/10.1007/s00122-016-2761-8

Porceddu A, Albertini E, Barcaccia G et al (2002) Linkage mapping in apomictic and sexual Kentucky bluegrass (Poa pratensis L.) genotypes using a two way pseudo-testcross strategy based on AFLP and SAMPL markers. Theor Appl Genet 104:273–280

Ripol MI, Churchill GA, Da SJAG, Sorrells M (1999) Statistical aspects of genetic mapping in autopolyploids. Gene 235:31–41

Rosyara UR, Bink MCAM, van de Weg E et al (2013) Fruit size QTL identification and the prediction of parental QTL genotypes and breeding values in multiple pedigreed populations of sweet cherry. Mol Breeding 32:875–887

Sasai R, Tabuchi H, Kishimoto K et al (2019) Development of molecular markers associated with resistance to meloidogyne incognita by performing quantitative trait locus analysis and genome-wide association study in sweetpotato. DNA Res 26:399–409. https://doi.org/10.1093/dnares/dsz018

Schmitz Carley CA, Coombs JJ, Douches DS et al (2017) Automated tetraploid genotype calling by hierarchical clustering. Theor Appl Genet 1–10. https://doi.org/10.1007/s00122-016-2845-5

Serang O, Mollinari M, Garcia AAF (2012) Efficient exact maximum a posteriori computation for Bayesian snp genotyping in polyploids. PLoS ONE 7:e30906. https://doi.org/10.1371/journal.pone.0030906

Shirasawa K, Tanaka M, Takahata Y et al (2017) A high-density SNP genetic map consisting of a complete set of homologous groups in autohexaploid sweetpotato (Ipomoea batatas). Sci Rep 7:44207. https://doi.org/10.1038/srep44207

Soares NR, Mollinari M, Oliveira GK et al (2021) Meiosis in polyploids and implications for genetic mapping: a review. Genes (basel) 12:1517. https://doi.org/10.3390/genes12101517

Soltis DE, Soltis PS (1993) Molecular data and the dynamic nature of polyploidy molecular data and the dynamic nature of polyploidy. Crit Rev Plant Sci 12:243–273

Soltis PS, Soltis DE (2012) Polyploidy and genome evolution. Springer, Berlin Heidelberg

Song Q, Yan L, Quigley C et al (2017) Genetic characterization of the soybean nested association mapping population. Plant Genome 10:plantgenome2016–10

Sorrells ME (1992) Development and application of RFLPs in polyploids. Crop Sci 32:1086–1091. https://doi.org/10.2135/cropsci1992.0011183X003200050003x

Stam P (1993) Construction of integrated genetic linkage maps by means of a new computer package: joinmap. Plant J 3:739–744. https://doi.org/10.1046/j.1365-313x.1993.03050739.x

Sturtevant AH (1913) The linear arrangement of six sex-linked factors in drosophila, as shown by their mode of association. J Exp Zool 14:43–59. https://doi.org/10.1002/jez.1400140104

Sybenga J (1975) Meiotic configurations. Springer

Taniguti CH, Taniguti LM, Amadeu RR et al (2022) Developing best practices for genotyping-by-sequencing analysis in the construction of linkage maps. Gigascience 12:giad092. https://doi.org/10.1093/gigascience/giad092

Thompson E (1984) Information gain in joint linkage analysis. MA J Math Appl Med Biol 1:31–49

Thompson PG, Hong LL, Kittipat Ukoskit, Zhu Z (1997) Genetic linkage of randomly amplified polymorphic Dna (RAPD) markers in sweetpotato. J Am Soc Hortic Sci 122:79–82. https://doi.org/10.21273/jashs.122.1.79

Ukoskit K, Thompson P (1997) Autopolyploidy versus allopolyploidy and low-density randomly amplified polymorphic DNA linkage maps of sweetpotato. J Am Soc Agric Sci 122:822–828

Vigna BBZ, Santos JCS, Jungmann L et al (2016) Evidence of allopolyploidy in urochloa humidicola based on cytological analysis and genetic linkage mapping. PLoS ONE 11:e0153764. https://doi.org/10.1371/journal.pone.0153764

Voorrips RE, Gort G, Vosman B (2011) Genotype calling in tetraploid species from bi-allelic marker data using mixture models. BMC Bioinformatics 12:172. https://doi.org/10.1186/1471-2105-12-172

Wadl PA, Olukolu BA, Branham SE et al (2018) Genetic diversity and population structure of the usda sweetpotato (Ipomoea batatas) germplasm collections using Gbspoly. Front Plant Sci 9. https://doi.org/10.3389/fpls.2018.01166

De Winton D, Haldane JBS (1931) Linkage in the tetraploidPrimula sinensis. J Genet 24:121–144. https://doi.org/10.1007/BF03020826

Wright S (1938) The distribution of gene frequencies in populations of polyploids. Proc Natl Acad Sci USA 24:372–377

Wu R, Ma C-X (2005) A general framework for statistical linkage analysis in multivalent tetraploids. Genetics 170:899–907. https://doi.org/10.1534/genetics.104.035816

Wu KK, Burnquist W, Sorrells ME et al (1992) The detection and estimation of linkage in polyploids using single-dose restriction fragments. Theor Appl Genet 83:294–300. https://doi.org/10.1007/bf00224274

Wu SS, Wu R, Ma C-X et al (2001b) A multivalent pairing model of linkage analysis in autotetraploids. Genetics 159:1339–1350

Wu R, Ma C-X, Wu SS, Zeng Z-B (2002b) Linkage mapping of sex-specific differences. Genet. Res. 79:85–96 https://doi.org/10.1017/s0016672301005389

Wu R, Ma C-X, Painter I, Zeng Z-B (2002a) Simultaneous maximum likelihood estimation of linkage and linkage phases in outcrossing

species. Theor Popul Biol 61:349–363. https://doi.org/10.1006/tpbi.2002.1577

Wu R, Ma CX, Casella G (2004) A bivalent polyploid model for mapping quantitative trait loci in outcrossing tetraploids. Genetics 166:581–595. https://doi.org/10.1534/genetics.166.1.581

Wu S, Lau KH, Cao Q et al (2018) Genome sequences of two diploid wild relatives of cultivated sweetpotato reveal targets for genetic improvement. Nat Commun 9:4580. https://doi.org/10.1038/s41467-018-06983-8

Wu R, Gallo-meagher M, Littell RC, Zeng Z (2001a) A general polyploid model for analyzing gene segregation in outcrossing tetraploid species. Genetics 159:869–882. https://doi.org/10.1093/genetics/159.2.869

Xie C, Xu S (2000) Mapping quantitative trait loci in tetraploid populations. Genet Res 76:105–115

Yan H, Ma M, Ahmad MQ et al (2022) High-density single nucleotide polymorphisms genetic map construction and quantitative trait locus mapping of color-related traits of purple sweet potato [ipomoea Batatas (L.) Lam.]. Front Plant Sci 12. https://doi.org/10.3389/fpls.2021.797041

Yang S, Chen S, Geng XX et al (2016) The first genetic map of a synthesized allohexaploid Brassica with A, B and C genomes based on simple sequence repeat markers. Theor Appl Genet 129:689–701. https://doi.org/10.1007/s00122-015-2657-z

Yu KF, Pauls KP (1993) Segregation of random amplified polymorphic DNA markers and strategies for molecular mapping in tetraploid alfalfa. Genome 36:844–851. https://doi.org/10.1139/g93-112

Yu J, Holland JB, McMullen MD, Buckler ES (2008) Genetic design and statistical power of nested association mapping in maize. Genetics 178:539–551

Zhao N, Yu X, Jie Q et al (2013) A genetic linkage map based on AFLP and SSR markers and mapping of QTL for dry-matter content in sweetpotato. Mol Breeding 32:807–820. https://doi.org/10.1007/s11032-013-9908-y

Zheng C, Voorrips RE, Jansen J et al (2016) Probabilistic multilocus haplotype reconstruction in outcrossing tetraploids. Genetics 203:119–131. https://doi.org/10.1534/genetics.115.185579

Zheng C, Jiang Z, Meng Y et al (2023) Construction of a high-density ssr genetic linkage map and identification of qtl for storage-root yield and dry-matter content in sweetpotato. Crop J 11:963–967. https://doi.org/10.1016/j.cj.2022.11.003

Zheng C, Amadeu RR, Munoz PR, Endelman JB (2021a) Haplotype reconstruction in connected tetraploid F1 populations. Genetics. https://doi.org/10.1093/genetics/iyab106

Zielinski M-L, Mittelsten Scheid O (2012) Meiosis in polyploid plants. Polyploidy and genome evolution. Springer, Berlin Heidelberg, Berlin, Heidelberg, pp 33–55

New Analytical Tools for Molecular Mapping of Quantitative Trait Loci in Sweetpotato

6

Guilherme da Silva Pereira, Carla Cristina da Silva, João Ricardo Bachega Feijó Rosa, Olusegun Olusesan Sobowale, Gabriel de Siqueira Gesteira, Marcelo Mollinari, and Zhao-Bang Zeng

Abstract

Quantitative trait loci (QTL) mapping is an important tool in sweetpotato research, contributing to the understanding of genetic architecture of various traits, including dry matter, nematode resistance, and flesh color. Early QTL work was carried out by using marker information alone via single marker analysis (SMA), or based on parent-specific linkage map using interval mapping (IM), composite interval mapping (CIM), and multiple interval mapping (MIM). Initially developed for inbred diploid species populations, these methods did not fully consider the complex autopolyploid, outcrossing nature of sweetpotato. Technological and methodological advances made it possible to obtain integrated, fully phased genetic maps for the crop. A random-effect MIM approach that leverages identity-by-descent based on QTL genotype conditional probabilities has been employed since with increasing power and resolution. To illustrate QTL identification in sweetpotato, we used publicly available data from 'Beauregard' × 'Tanzania' full-sib family ($N = 315$) evaluated for flesh color in Peru. Several methods were able to detect two QTL on chromosomes 3 and 12 each for this trait in the same genomic regions. Despite the importance of such methods, there is need to extend existing models to account for multi-trait or multi-environment data and to evaluate their application in genomic-enabled prediction.

Keywords

Ipomoea batatas · QTL · Linkage map · Molecular markers · Marker-assisted selection

6.1 Introduction

Most agronomically important traits are controlled by many regions in the genome, which makes traits targeted by breeding programs usually quantitative in nature and more or less influenced by environment. The breeders need to evaluate a lot of candidate genotypes in multiple locations and years to select the best ones according to the product profiles they have in place. This process takes a lot of time and resources, making opportunities

G. da Silva Pereira (✉) · C. C. da Silva · J. R. B. F. Rosa · O. O. Sobowale
Federal University of Viçosa, Viçosa, Brazil
e-mail: g.pereira@ufv.br

G. de Siqueira Gesteira · M. Mollinari · Z.-B. Zeng
Bioinformatics Research Center,
Department of Horticultural Science, North Carolina State University, Raleigh, NC 27695, USA

G. C. Yencho et al. (eds.), *The Sweetpotato Genome*, Compendium of Plant Genomes,
https://doi.org/10.1007/978-3-031-65003-1_6

for variety renewal relatively cumbersome. For a crop species, product profiles vary according to countries and breeding programs within countries. In the case of sweetpotato, product profiles might include flesh and skin colors, β-carotene, sugar and dry matter contents, resistance to pests and diseases, and high yield, among others (Lindqvist-Kreuze et al. 2023). One way to accelerate this process is to use DNA-based markers genetically associated with quantitative traits to perform marker-assisted selection (MAS).

To implement MAS, one first initiative is to find which genomic regions underlie the variation of traits of interest (Collard and Mackill 2007). These regions are called quantitative trait loci (QTL) and the process of finding association between the genotype (molecular markers) and the phenotype (trait expression) is called QTL mapping. By doing so, we aim to describe the genetic architecture of quantitative traits of interest, i.e., we seek to uncover the number of loci influencing the variation in such traits, along with their respective map (or genome) locations and effects. QTL mapping studies follow a typical workflow, consisting of the (i) choice of parental varieties and obtaining the segregation population, (ii) collection of phenotypic data from such population, (iii) genotypic evaluation with polymorphic molecular markers, (iv) linkage map construction, and (v) QTL mapping itself. Here, we will quickly describe the first four requirements, but will mostly focus on previous and current methods for conducting QTL mapping in sweetpotato.

Despite our somewhat historical perspective on QTL methods, this chapter has no intention of covering all aspects and details involved in such models, but rather providing basic understanding of QTL methods specifically applied in sweetpotato research.

6.2   QTL Mapping

A QTL mapping study starts with the need to generate a biparental mapping population that segregates for the trait(s) of interest. On the one

hand, if pure lines were available for sweetpotato, backcross, F_2 and recombinant inbred line (RIL) populations could easily be employed. On the other hand, when dealing with highly heterozygous sweetpotato parents, full-sib families (i.e., segregating F_1 populations) are generally used. Some of the reasons for absence of inbred lines in outcrossing species were mentioned in previous chapters. Self-incompatibility, inbreeding depression, and its autopolyploid nature are among the major factors preventing the existence of sweetpotato pure lines.

For outcrossing species in general, F_1 populations can be utilized for linkage map construction and QTL analysis. If both parents are phenotypically contrasting due to complete fixation of respective Q and q alleles of a given QTL (e.g., $QQQQQQ \times qqqqqq$), there will be no segregation for that locus as all F_1 individuals will be $QQQqqq$. If it represents a major QTL, less (heritable) variation will be noticed in the progeny and marginal variation due to minor QTL will be hardly detectable, especially under limited population sizes. For most diploid linkage-based QTL studies providing sufficient resolution and statistical power, population sizes greater than 200 individuals are often utilized in literature. Though such studies are not very common in hexaploid species, based on experience, population sizes greater than 250–300 individuals are recommended.

Given that the parents have been crossed and the population has been established in a screenhouse, sweetpotato progenies can be cloned to constitute plant material for genotyping as well as for phenotyping trials. Such trials can be conducted in screenhouse or field depending on which traits will be evaluated. Reaction to disease due to artificial infection or drought-related traits are more easily assessed in screenhouse trials where environmental control is usually better, for example, whereas most of the other traits can be accessed via field trials. In any case, experimental designs must be employed to increase the accuracy of individual mean estimates and, consequently, the ability to detect QTL. Making sure that the residual variation is kept to a minimal and therefore genetic variation can be dissected in QTL studies is imperative. In fact, because sweetpotato can be clonally propagated, plot trials can have more than one plant and experiments can be replicated in different locations, seasons, or conditions. Although mapping populations are created with certain trait(s) in mind, other varying traits are often studied for the same mapping population.

Genotyping is conducted based on DNA extracted from healthy, clean, fresh leaves from plant material available at screenhouse or field. Methods for obtaining molecular markers have been discussed in previous chapters. Here, we will just reinforce that advancing of genotyping platforms, such as those based on next-generation sequencing (NGS), allow for variation detection at the single base resolution level for thousands of markers, namely single nucleotide polymorphisms (SNPs). In addition, the number of reads from each SNP can be leveraged for allele dosage or micro-haplotyping purposes, increasing the informativeness of such markers for linkage and subsequent QTL mapping (Hackett et al. 2014; Mollinari et al. 2020). Previous genotyping platforms, such as those based on electrophoretic gels, serve for lots of purposes in genetic studies, but they are particularly limiting in the case of polyploids. The number of confounded classes using dominant markers increases from 2 in diploids ($0 = aa$, versus $1 = Aa$, or AA), to 4 classes in autotetraploids ($0 = aaaa$ versus $1 = Aaaa$, $AAaa$, $AAAa$, or $AAAA$), and to 6 classes in autohexaploids ($0 = aaaaaa$ versus $1 = Auaaaa$, $AAaaaa$, $AAAaaa$, $AAAAaa$, $AAAAAa$, or $AAAAAA$).

As described in the previous chapter, linkage mapping is then conducted to group, order and phase such markers. In the context of where there is missing data—as it is, in fact, the case of allele dosage-based SNPs—, methods using hidden Markov models should be preferred (Mollinari et al. 2020). The ultimate goal of linkage mapping is to provide a comprehensive view of segregation of homologous chromosomes from parents to progeny and, by doing so, to allow the computation of possible QTL

genotypes—including those between marker intervals—that each individual is most likely to carry. The computation of QTL genotype probabilities conditional to a map is the basis of the most employed QTL detection methods that we will describe here. Even in the context of high-density maps, fully phased maps help with figuring out the best haplotype(s) to be targeted for MAS (Gemenet et al. 2020a).

Finally, QTL mapping can be performed using different statistical genetics methods such as single marker analysis—SMA (Stuber et al. 1987; Edwards et al. 1987), interval mapping—IM (Lander and Botstein 1989), composite interval mapping—CIM (Jansen and Stam 1994; Zeng 1994), and multiple interval mapping—MIM (Kao et al. 1999). Except for SMA that relies on the marker information alone, all other methods employ the marker information in the context of a linkage map. All these methods were initially developed and broadly used in diploid species where traditional populations such as backcross, F_2 or RIL were available. Before integrated, fully phased maps were available for sweetpotato, most of these methods have been applied to previous studies of the crop. However, in most cases, IM and its variations were employed using separate maps, one for each parent, when computing QTL genotype conditional probabilities (Cervantes-Flores et al. 2008b). Currently, QTL mapping based on integrated maps can be performed for complex autopolyploid species (Da Silva Pereira et al. 2020). A summary of published QTL studies in the crop so far is available in Table 6.1.

We will take advantage of previously published data from the 'Beauregard' x 'Tanzania' (BT) population (Gemenet et al. 2020a) as an example to illustrate QTL identification in sweetpotato through a range of methods, discussing its basis but without going into much technical details. 'Beauregard' is an orange-fleshed American variety, with low dry matter content and susceptible to nematodes (namely *Meloidogyne incognita* and *M. enterolobii*) and sweetpotato virus disease (SPVD) (Rolston et al. 1987), whereas 'Tanzania' is a

Table 6.1 Quantitative trait loci(QTL) mapping studies for sweetpotato

Mapping population	Population size	Marker system	Trait	Method	Number of QTL	PEV (%)	Software	References
Tanzania × Beauregard	240	AFLP	Root-knot nematode	SMA, IM, CIM	9	2.2–11.5	WinQTL Cartographer	Cervantes-Flores et al. (2008b)
Nancy Hall × Tainung 27	120	ISSR	Yield-related: several traits	CIM	24	14.1–29.9	WinQTL Cartographer	Chang et al. (2009)
Tanzania × Beauregard	240	AFLP	Dry matter, starch, β-carotene	CIM	13, 12, 8	15.0–24.0, 17.0–30.0, 17.0–35.0	WinQTL Cartographer	Cervantes-Flores et al. (2011)
Xushu 18 × Xu 781	202	AFLP, SSR	Dry matter	CIM	27	9.0–45.1	MapQTL v4.0	Zhao et al. (2013)
Luoxushu 8 × Zhengshu 20	240	SRAP	Storage root yield	IM, CIM	45	10.2–59.3	MapQTL v4.0	Li et al. (2014)
Xushu 18 × Xu 781	202	AFLP, SSR	Starch	IM, CIM	8	9.1–38.8	MapQTL v4.0	Yu et al. (2014)
New Kawogo × Beauregard	240	SRR	Dry matter, starch, β-carotene	SMA	4, 6, 8	3.1–4.4, 4.3–6.9, 2.0–7.4	PROC GLIMMIX (SAS)	Yada et al. (2017)
Beauregard × Tanzania	315	SNP	Yield-related traits	MIM	13	8.9–22.0	QTLpoly	Da Silva Pereira et al. (2020)
Beauregard × Tanzania	315	SNP	Dry matter, starch, β-carotene, flesh color	MIM	5, 2, 2, 5	6.0–37.4, 17.0–51.2, 29.0–50.2, 3.2–53.6	QTLpoly	Gemenet et al. (2020a)
Jizishu 1 × Longshu 9	300	SSR	Root rot	SMA, IM, MIM	5	52.6–57.0	SPSS Statistics, MapQTL v5.0	Ma et al. (2020)
Beauregard × Tanzania	315	SNP	β-carotene, iron, zinc	MIM	2, 2, 2	26.5–39.3,19.0–32.0, 11.3–12.2	QTLpoly	Mwanga et al. (2021)
Tanzania × Beauregard	244	SNP	Root-knot nematode	MIM	1	58.3	QTLpoly	Oloka et al. (2021)
Xuzishu8 × Meiguohong	274	SLAF-seq	Flesh color, skin color, anthocyanin content	CIM	1, 1, 8	36.3, 45.9, 10.5–28.5	R/qtl	Yan et al. (2022)
Xushu 18 × Xu 781	500	SSR	Storage root yield, dry matter content	SMA, IM, MIM	33, 16	6.5–47.5, 3.2–18.9	SPSS statistics	Zheng et al. (2023)

AFLP amplification fragment length polymorphism; *SSR* simple sequence repeats; *ISSR* inter SSR; *SRAP* sequence-related amplified polymorphism; *SNP* single nucleotide polymorphism; *SMA* single marker analysis; *IM* interval mapping; *CIM* composite interval mapping; *MIM* multiple interval mapping; *PEV* proportion of variance explained by QTL cream-fleshed African landrace showing high dry matter content and resistance to nematodes and SPVD (Mwanga et al. 2001). This population was obtained and evaluated in five environments in Peru for several traits, including flesh color which is our target here (Fig. 6.1a). Flesh color was evaluated for 315 progenies based on scores ranging from 1 (white) to 8 (dark orange) (Grüneberg et al. 2019), and adjusted means obtained as described before (Gemenet et al. 2020a) (Fig. 6.1b). The population was genotyped using a quantitative genotyping-by-sequencing based protocol (GBSpoly) and the reads were aligned against both *Ipomoea trifida* and *I. triloba* genomes. A total of 38,701 SNPs were used for map construction, and 17 progenies have been filtered out, making up to a population size of 298 (Mollinari et al. 2020).

6.2.1 Single-Marker Analysis

Single-marker analysis (SMA) can be carried out using any statistical method that tests whether the differences among mean classes are significant or not, such as *t*-tests or analysis of variance (ANOVA) derived F-tests. For example, if we are interested in testing (additive) effects of single markers in a diploid F_2 population derived from inbred parents, genotypic classes of codominant molecular markers can be scored as 0 (*aa*), 1 (*Aa*), or 2 (*AA*) depending on the number—or dosage—of a certain alternate allele *A*. Using the same reasoning, hexaploid genotypic classes can be represented by 0 (*aaaaaa*), 1 (*Aaaaaa*), …, up to 6 (*AAAAAA*). A simple linear regression model relating y_i, the phenotype of individual *i* (or the response variable), to x_i, the genotype of individual *i* (or the explanatory variable), can be performed as follows:

$$y_i = \mu + \beta x_i + \varepsilon_i$$

where μ is the intercept; β is the regression coefficient representing the expected change in y_i for a one-unit change in x_i or, in other words, the additive effect as the average effect of allele substitution (when an *a* is replaced by an *A*); and $\varepsilon_i \sim N(0, \sigma^2)$ is the residual term, expected to be normally distributed with mean zero and variance σ^2. The residual term is where all the unexplained variation of variable *y* goes after fitting variable *x*. Fitting this model by ordinary least squares, the coefficient of determination,

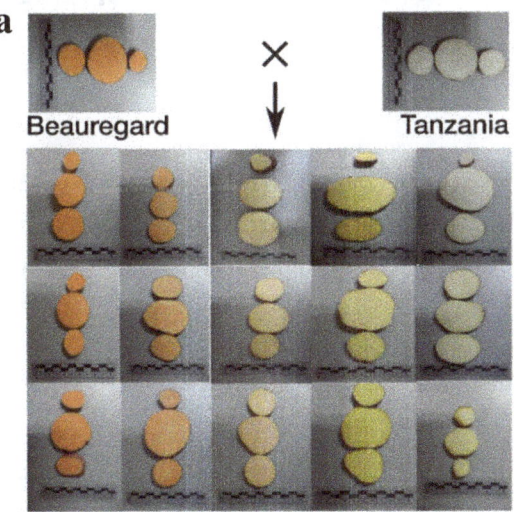

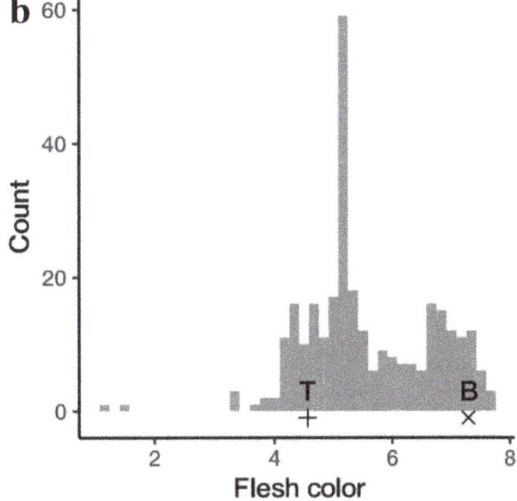

Figure 6.1 Phenotypic segregation for flesh color in the 'Beauregard' × 'Tanzania' full-sib population (*N* = 315) evaluated in Peru. **a** Each photo depicts a different progeny (bottom) in comparison to their parents (top). **b** Distribution of adjusted means along with parents 'Beauregard' (B) and 'Tanzania' (T). Adapted from Gemenet et al. (2020a)

R^2, equals one minus the ratio between residual sum of squares and total sum of squares. This estimate is often interpreted as the proportion of variance explained (*PEV*) by marker (or QTL) in the context of QTL analysis.

Maximum likelihood can be leveraged for parameter estimation and assessing significance. In our case, we are interested in knowing which hypothesis, the null ($H_0 : \beta = 0$) or alternate one ($H_1 : \beta \neq 0$), is to be rejected given the data. In this case, the likelihood $\mathcal{L}_1$ of full model (including variable x, thus under H_1) is tested against the likelihood $\mathcal{L}_0$ of reduced model ($y_i = \mu + \varepsilon_i$, thus under H_0) by means likelihood ratio test (*LRT*) as follows:

$$LRT = 2 \times \log \frac{\mathcal{L}_1\left(\mu, \beta, \sigma^2\right)}{\mathcal{L}_0\left(\mu, \sigma^2\right)}$$

LRT is assumed to have a chi-squared distribution, with degrees of freedom equals the number of classes minus one, from which *P*-values can be drawn. Genome-wide threshold for declaring significant QTL can be obtained through permutation tests (Doerge and Churchill 1996).

In our illustration, a total of 28,651 SNPs derived from alignment to *I. trifida* reference genome has been tested using an additive (dosage) model (Fig. 6.2a). Several models could have been tested in order to try to find associations between phenotype and genotype in autopolyploids, like those proposed at GWASpoly R package (Rosyara et al. 2016). In fact, SMA consists of a typical Genome Wide Association Study (GWAS) model without the need to controlling for population structure or cryptic relatedness. The SNP

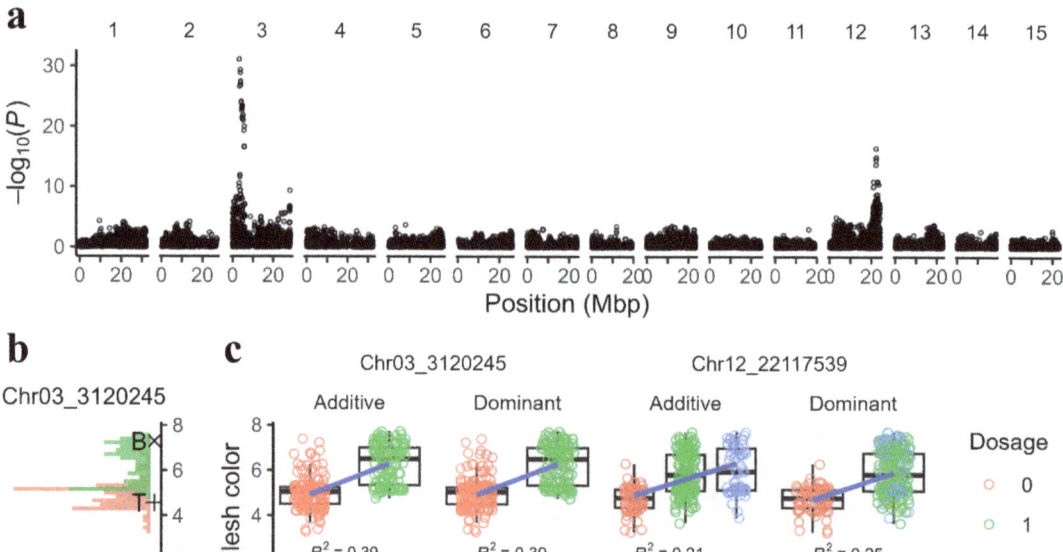

Fig. 6.2 Single-marker analysis for flesh color in sweetpotato 'Beauregard' × 'Tanzania' full-sib population ($N = 298$) evaluated in Peru. **a** Log of *P*-values for simple linear regression for 28,651 SNPs based on an additive model. Markers are ordered according to *Ipomoea trifida* genome. **b** Phenotype distribution (along with parents B and T) according to SNP Chr03_2120245 dosage classes. **c** Simple linear regression of the top associated markers on chromosome 3 (3,120,245 bp) and 12 (22,117,539 bp) depicting additive and dominant models. Dominant models mean dosages >0 are grouped into a single genotype class 1 (i.e., at least one *A*)

Chr03_3120245 (Fig. 6.2b) contributed most ($PEV = 39\%$) to explaining flesh color variation in BT population, followed by Chr12_22117539 ($PEV = 21\%$).

The additive model tests the linear relationship between flesh color scores and allele dosage data as explained earlier, thus we expect to see scores increasing (positive slope) or decreasing (negative slope) if they are significantly associated with a marker. Depending on how many dosage classes one marker shows, dominant models will be equivalent to the additive one. Examples of dominant models, like what electrophoresis gel-based markers would only allow testing for, are shown for the same markers for comparison purposes. In the case of Chr12_22117539 (three classes), $PEV = 25\%$ for the dominant model whereas in the case of Chr03_3120245 (two classes), the two models are equivalent (Fig. 6.2c). For highly heritable traits, such as flesh color ($H^2 = 0.92$ in our example) and high-density genotyping individuals, SMA might be able to detect genomic regions associated with such traits as seen here.

Even less informative markers, such as those based on amplification fragment length polymorphism (AFLP), were first used to carry out SMA by Cervantes-Flores et al. (2008b) when working with reciprocal 'Tanzania' × 'Beauregard' (TB) population ($N = 240$). They found nine markers associated to root-knot nematode (RKN) resistance, where both 'Tanzania' (seven) and 'Beauregard' (two) appeared to hold resistance alleles, ranging from 11.5% to 2.2% of the total variation. Similarly, Yada et al. (2017) using 'New Kawogo' × 'Beauregard' (NKB) population ($N = 240$) (Yada et al. 2015) were able to identify 12, 4, 6, and 8 SSR markers associated to yield ($PEV = 4.2 \sim 9.1\%$), dry matter ($PEV = 3.1 \sim 4.4\%$), starch ($PEV = 4.3 \sim 6.9\%$), and β-carotene ($PEV = 2.0 \sim 7.4\%$). As observed, despite limitations in gathering high-density markers at that time, there has been progress in characterizing variation of important traits in sweetpotato via molecular markers when genetic maps were not available.

6.2.2 Fixed-Effect Interval Mapping Model

Interval mapping (IM) was first introduced in the context of newly developed linkage maps based on multipoint estimations using the hidden Markov model framework (Lander and Botstein 1989). Such maps are used for the computation of conditional probability distribution of genotypes (Jiang and Zeng 1997) allowing for a systematic search of QTL, including within marker intervals (inter-marker search). This idea, initially implemented for diploid inbred-derived populations, was later extended to accommodate both diploid (Wu et al. 2002) and autopolyploid (Mollinari and Garcia 2019) outbred-derived populations. In any case, because we do not observe the QTL genotypes, they are treated as latent variables and can be modeled as a mixture of normal distributions. An F_2 model for testing QTL additive effects, e.g. every 1 cM, can be represented as follows:

$$y_i = \mu + \beta^* x_i^* + \varepsilon_i$$

where x_i^* is an indicator variable with probabilities of individual i being 0 (qq), 1 (Qq) or 2 (QQ) at given position; β^* represents the additive effect of a QTL (instead of a marker); and $\varepsilon_i \sim N(0, \sigma^2)$. Again, LRT can be carried out based on the ratio between likelihoods of models under alternate hypothesis $\mathcal{L}_1(\mu, \beta^*, \sigma^2)$ and null hypothesis $\mathcal{L}_0(\mu, \sigma^2)$, known as odds ratio. Broadly preferred for interpretation (and plotting) purposes, logarithm of the odds ratio (LOD) scores can be obtained by using:

$$LOD = \log_{10} \frac{\mathcal{L}_1(\mu, \beta^*, \sigma^2)}{\mathcal{L}_0(\mu, \sigma^2)}$$

Or simply by using $LOD = LRT / [2 \times \log(10)]$.

An extension of IM, composite interval mapping (CIM), proposes the inclusion of M markers as covariates (also called cofactors) in order to control variation outside the region being search for QTL, increasing the detection power (Zeng 1994), as follows:

$$y_i = \mu + \beta^* x_i^* + \sum_{m=1}^{M} \beta_m x_{mi} + \varepsilon_i$$

Similar to what can be done for SMA, in order to evaluate significance (declare a QTL), empirical *LOD* thresholds are computed for each trait using permutations (Doerge and Churchill 1996). Such models and algorithms for running IM and CIM are available in software like WinQTL Cartographer (Basten et al. 1999) and MapQTL (van Ooien et al. 2000), both broadly used in sweetpotato QTL mapping work (Table 6.1).

This model can be employed in linkage maps constructed using the double pseudo-test cross method, resulting in two separate maps, one for each parent (Grattapaglia and Sederoff 1994). After building such maps for TB population (Cervantes-Flores et al. 2008a), IM and CIM was used for QTL confirmation for RKN resistance (Cervantes-Flores et al. 2008b) as well as for QTL identification for dry matter (13, *PEV* = 15 ∼ 24%), starch (12, *PEV* = 17 ∼ 30%), and β-carotene (8, *PEV* = 17 ∼ 35%) in the TB population (Cervantes-Flores et al. 2011). Both methods were also used to identify other 27 QTL in different environmental conditions for dry matter (*PEV* = 9.0 ∼ 45.1%) (Zhao et al. 2013), and 8 QTL for starch (*PEV* = 9.1 ∼ 38.8%) (Yu et al. 2014). For yield traits, 23 QTL have identified using CIM in 'Nancy Hall' (*PEV* = 14.1 ∼ 29.8%) and 'Tainung 27' (*PEV* = 16.0 ∼ 29.9%) separate maps, respectively (Chang et al. 2009), whereas another study identified 45 QTL using IM and CIM, explaining between 10.2 and 59.3% of the phenotypic variation.

A first approach to map QTL in autopolyploid species using the information of fully phased linkage maps has been initially proposed for autotetraploids (Hackett et al. 2014), herein called fixed-effect interval mapping (FEIM) model. It consists of a single-QTL model, where every position is tested according to a model that can be more generally written for any given ploidy *p* as follows:

$$Y = \mu_C + \sum_{j=2}^{p} \alpha_j X_j + \sum_{j=p+2}^{2p} \alpha_j X_j$$

where μ_C is the intercept, and α_j and X_j are the main additive effects and indicator variables for allele *j* (i.e., haplotype probabilities inferred from fully phased linkage maps), respectively, where $j = \{1, \ldots, p\}$ and $j = \{p + 1, \ldots, 2p\}$ represent the two sets of alleles, one for each parent. The constraints $\alpha_1 = 0$ and $\alpha_{p+1} = 0$ are imposed to satisfy the conditions $\sum_{i=1}^{p} X_j = p/2$ and $\sum_{i=p+1}^{2p} X_j = p/2$, so that μ_C is a constant hard to interpret due to these constraints. Notice that the higher the ploidy level, the more effects must be estimated. For example, tetraploid models have six main effects, hexaploid models have 10 effects, octoploid models will have 14 effects (i.e., $2p - 2$), which will be needed for every new QTL added in a multiple loci model.

Such a model has been implemented for hexaploid species in R packages like polymapR (Bourke et al. 2018) and QTLpoly (Da Silva Pereira et al. 2020). Application of QTLpoly function 'feim()' in our flesh color illustration shows two QTL in the same genomic regions as identified using SMA (Fig. 6.3a). The analysis used the linkage map reported for the population, combining a total of 38,701 SNPs aligned against both *I. trifida* and *I. triloba* genomes (Mollinari et al. 2020). The QTL on chromosome 3 at 34.11 cM (*LOD* = 31.21, *PEV* = 41.8%) has its peak close to SNP Chr03_2615608, whereas QTL on chromosome 12 at 146.02 cM (*LOD* = 17.98, *PEV* = 21.6%) was mapped close to SNP Chr12_22131994 (both in relation to *I. trifida* reference genome). LOD threshold for 95% genome-wide significance equals 7.7 was obtained after 1,000 permutation tests (Fig. 6.3b).

6.2.3 Multiple QTL Random-Effect Model

Although all these studies were important and made progress in understanding agronomically important traits in sweetpotato, they were

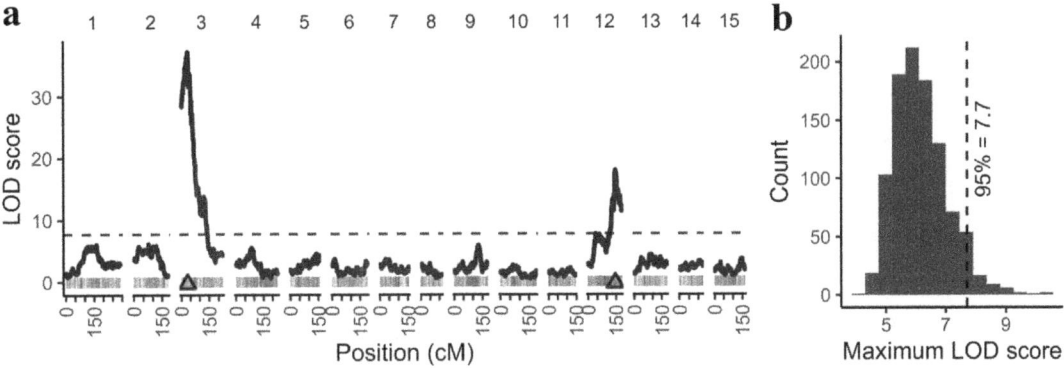

Fig. 6.3 Fixed-effect interval mapping analysis of flesh color in sweetpotato 'Beauregard' × 'Tanzania' full-sib population ($N = 298$) evaluated in Peru. **a** Log-of-the-odds (LOD) score profile showing QTL on chromosomes 3 and 12 (triangles). **b** Distribution maximum LOD scores from 1000 permutation tests and genome-wide significance threshold for $\alpha = 0.05$

relatively limited by the QTL mapping method. In fact, IM and CIM offer improvements in comparison to SMA by allowing marker interval testing and increased detection power. However, they all consisted of single-QTL models. One of our expectations when conducting QTL mapping studies is realizing that multiple QTL, in fact, contribute toward the trait variation. In this scenario, multiple interval mapping (MIM) model together with an algorithm for searching QTL was needed, similar to what diploid, inbred-based populations had (Kao et al. 1999).

Our method is based on the following random-effect model (Da Silva Pereira et al. 2020):

$$y = 1\mu + \sum_{q=1}^{Q} g_q + \varepsilon$$

where the vector of phenotypic values from a specific trait y is a function of the fixed intercept μ, the $q = 1, \ldots, Q$ random QTL effects $g_q \sim N\left(0, G_q\sigma_q^2\right)$, and the random environmental error $\varepsilon \sim N\left(0, I\sigma^2\right)$. G_q is an identity-by-descent (IBD) matrix comparing all possible 400 genotypes in an autohexaploid biparental population (i.e., whether two individuals share from 0 to 6 alleles IBD) according to genotype conditional probabilities of QTL q, working similarly to an additive relationship matrix. Because we only need to estimate one parameter per QTL (the very variance component associated with

it), it is relatively easy to look for additional QTL and add them to the variance component model, without ending up with an overparameterized model.

A multiple-QTL model is known to have increased power when compared to a single-QTL model, with ability to detect minor or separate yet linked QTL (Da Silva Pereira et al. 2020). Variance components associated with putative QTL (σ_q^2) are tested using score statistics from the R package varComp (v. 0.2–0) (Qu et al. 2013). Final models are fitted using residual maximum likelihood (REML) from the R package sommer (v. 3.6) (Covarrubias-Pazaran 2016). Rather than guessing pointwise significance levels for declaring QTL, we use the score-based resampling method to assess the genome-wide significance level α (Zou et al. 2004).

Building a multiple QTL model is considered a model selection problem, and there are several ways to approach it. QTLpoly tries to provide functions flexible enough, so that the users can build a multiple QTL model on their own, manually. The strategy mentioned below has been tested through simulations and it is implemented in a function called 'remim()'. It consists of an adaptation of the algorithm proposed by Kao et al. (1999) for fixed-effect MIM for diploids to our random-effect MIM (REMIM) for polyploids, which is summarized as follows:

1. Null model. For each trait, a model starts with no QTL:

$$y = 1\mu + \varepsilon$$

2. Forward search. QTL $(q = 1, \ldots, Q)$ are added one at a time, conditional to the one(s) (if any) already in the model, under a less stringent genome-wide significance level (e.g., $\alpha < 0.20$):

$$y = 1\mu + \sum_{q=1}^{Q} g_q + \varepsilon$$

3. Model optimization. Each QTL r is tested again conditional to the remaining one(s) in the model under a more stringent genome-wide significance level (e.g., $\alpha < 0.05$):

$$y = 1\mu + g_r + \sum_{q \neq r} g_q + \varepsilon$$

Steps 2 and 3 are repeated until no more QTL can be added to or dropped from the model, and positions of the remaining QTL do not change. After the first model optimization, any following forward searches use the more stringent threshold (e.g., $\alpha < 0.05$) as the detection power is expected to increase once QTL have already been added to the model.

4. QTL profiling. Score statistics for the whole genome are updated conditional to the final set of selected QTL. Once the final model is fitted, QTL heritability is computed as $h_q^2 = \sigma_q^2/\sigma_p^2$, where σ_p^2 is the total phenotypic variance.

The BT mapping population was leveraged to identify two major QTL related to starch, β-carotene, and their respective correlated traits, dry matter and flesh color (Gemenet et al. 2020a)—the same one described here in our example (Fig. 6.4a). Again, two QTL, one on chromosome 3 at 36.14 cM ($P < 10^{-16}$,

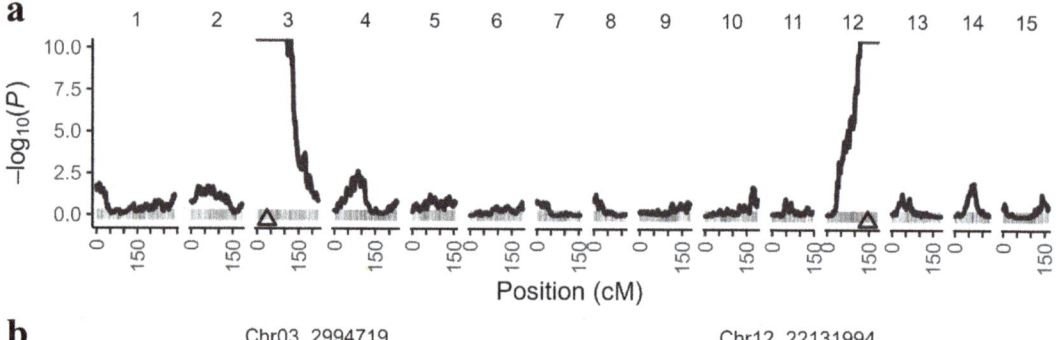

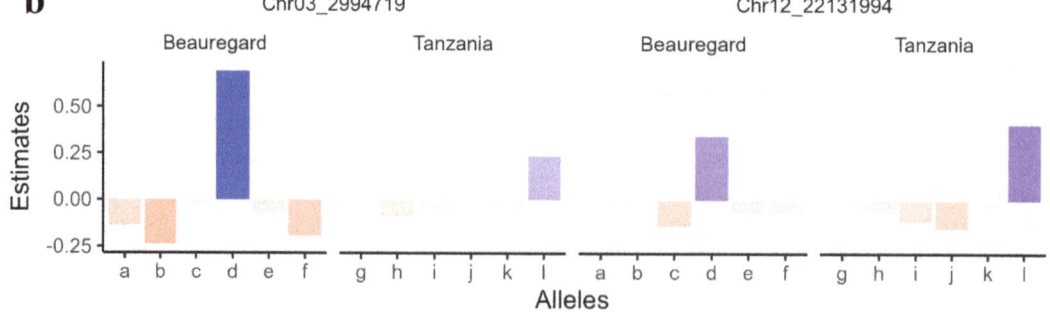

Fig. 6.4 Random-effect multiple interval mapping analysis for flesh color in sweetpotato 'Beauregard' × 'Tanzania' full-sib population ($N = 298$) evaluated in Peru. **a** Log of P-value profile showing QTL (triangles) on chromosomes 3 ($h_q^2 = 0.536$) and 12 ($h_q^2 = 0.293$). **b** Prediction of parental haplotype contributions to increasing (blue) or decreasing (red) overall mean $\mu = 5.68$ per QTL. Adapted from Gemenet et al. (2020a)

$h_q^2 = 0.536$) was located close to SNP Chr03_2994719 (thus close to FEIM results), and one on chromosome 12 at 146.02 cM ($P < 10^{-16}$, $h_q^2 = 0.293$) was mapped close to SNP Chr12_22131994 (same position as for FEIM). Together with RNA-seq data, this research has shown that the QTL on chromosome 3 presented a correlated effect in reducing starch (and dry matter) and increasing β-carotene contents (and flesh color scores, Fig. 6.4b) in genotypes carrying a haplotype from the 'Beauregard' parent, shedding light into the genetics basis of negatively correlated traits (dry matter and flesh color), very well known to breeders (Gemenet et al. 2020a).

Considering the same BT population integrated linkage map (Mollinari et al. 2020) and the REMIM approach, 13 QTL were mapped for eight yield-related traits, with the number of QTL per trait ranging from one to four. These QTL explained up to 55% of the total variation, where both parents ('Beauregard' and 'Tanzania') contributed with alleles to increasing the trait means (Da Silva Pereira et al. 2020). Studying iron (Fe) and zinc (Zn) contents in Ghana for BT population, two QTL each were found explaining respective 51.0%, and 23.5% of total variation, in the same location as those QTL for β-carotene (Mwanga et al. 2021), making double biofortification efforts likely to be successful. Finally, one major QTL explaining 58.3% of total variation for root-knot nematode resistance for the reciprocal TB population was also detected (Oloka et al. 2021).

We are currently carrying out validation tests on SNPs converted into kompetitive allele specific PCR (KASP) within QTL regions associated with several traits (Da Silva Pereira et al. 2023). In our flesh color illustration, the genome of *I. trifida* shows six transcripts annotated as *phytoene synthase*, namely itf12g01830.t1, itf03g05110.t1, itf03g10720.t1, itf03g10720.t2, itf14g07540.t1, itf14g07550.t1 (http://sweetpotato.uga.edu/). From the QTL mapping analysis, we found out that *itf03g05110* is the most likely gene involved in variation of flesh color in the BT population, as it matches the location

of QTL on chromosome 3. Polymorphic SNPs derived from whole-genome sequencing of the 16 parents of an 8×8 diallel called 'Mwanga Diversity Panel' (MDP) (Wu et al. 2018) were selected within *itf03g05110*. Results for one SNP, Chr03_3120259, are shown here. Allele dosage was estimated using fitPoly R package (Voorrips et al. 2011) (Fig. 6.5a), and tested against the flesh color scores of a sample of the 16 parents plus 78 progenies from MDP. The results have shown significant association ($P = 8.0 \times 10^{-6}$) between genotype and phenotype, with 21% of proportion of variance explained by this single marker (Fig. 6.5b), making it a candidate for MAS purposes in sweetpotato.

6.3 BSA-seq

As observed in the previous subsection, the detection of QTL depends on the availability of DNA molecular markers and the construction of linkage maps from biparental crosses with segregating phenotypes. As such, fine QTL mapping requires a high number of polymorphic markers and a large population size. Although sequencing technologies costs had lowered over time, high-throughput SNP genotyping of large populations is still costly, especially for polyploid species which require high sequencing depth to accurately perform dosage calling (Gemenet et al. 2020b).

Bulk-segregant analysis sequencing (BSA-seq) is a strategy that enables the identification of SNPs associated to traits of interest in a less expensive way when compared to conventional QTL mapping strategies, by combining bulked-segregant analysis (BSA) (Michelmore et al. 1991) and whole genome sequencing (WGS). The DNA of progenies in the extremes of the trait distribution are bulked according to their phenotypic class—one called 'low' and another 'high' bulks. Both bulks are subjected to WGS, and the resulting reads are mapped to a reference genome. A similar frequency distribution of alleles from both parents is expected in

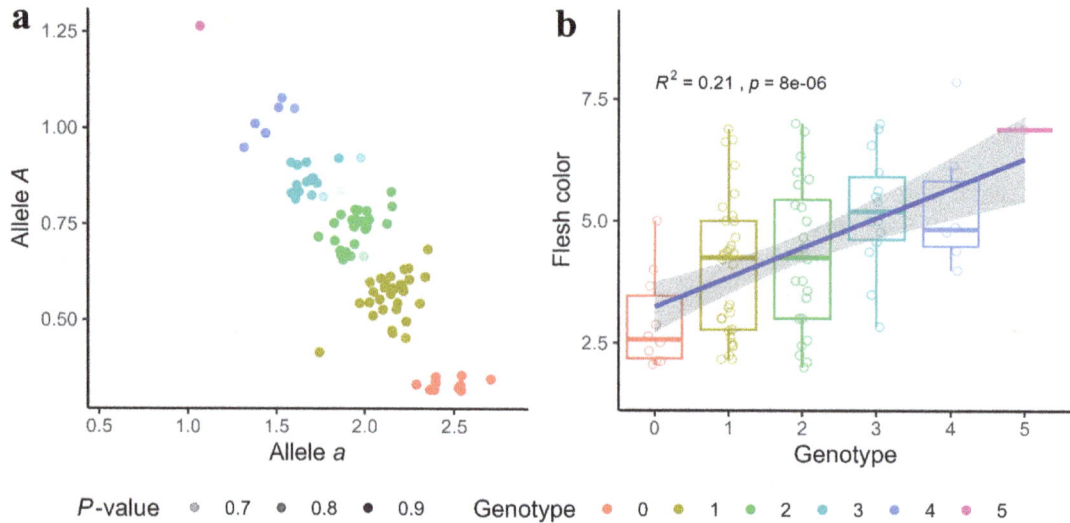

Fig. 6.5 QTL for flesh color on chromosome 3 converted into kompetitive allele specific PCR (KASP) marker. The marker was designed at 3,120,259 bp within *phytoene synthase* gene (*itf03g05110*) of *Ipomoea trifida* genome. **a** Dosage calling from alleles *a* and *A* intensities where each dot represents an individual assigned to given genotype class (color) under certain probability *P* (transparency). **b** Association with allele dosage shows an $R^2 = 0.21$ ($P = 8.0 \times 10^{-6}$) for samples of an 8 × 8 diallel ($N = 94$). Adapted from Da Silva Pereira et al. (2023)

regions that are not associated with the phenotype expression, while an uneven representation from one of the two parental alleles is expected in QTL regions.

This strategy was first proposed in yeast (Ehrenreich et al. 2010) and later applied to rice under so-called QTL-seq, where QTL for blast disease resistance and seedling height were successfully identified (Takagi et al. 2013). The proportion of reads derived from each parental genome was used to determine a SNP-index where 0 and 1 represent the entirety of reads containing the reference or alternative allele, respectively, and 0.5 represents equal contributions. The difference between indexes from both low and high bulks, called ΔSNP-index, will show values around 0 for the latter and close to 1 for the former case. Since then, BSA-seq has been used for several species such as tomato (Wen et al. 2019), capsicum (Park et al. 2019), and soybean (Zhang et al. 2018).

BSA-seq was first applied to a polyploid species in 2018. Clevenger et al. (2018) used the strategy to identify QTL for late leaf spot disease resistance in allotetraploid peanut. The initial hindrance was that the SNP detection methods used for diploid species produced a high proportion of false positive SNPs in peanuts. To circumvent this issue, the authors used a polyploid SNP calling pipeline. The polyploid calling allowed the identification of three QTL and the development of SNP markers for MAS. Recently, BSA-seq was used in combination with other techniques for QTL detection and MAS application for seed weight in peanut (Wang et al. 2022), and disease resistance in allotetraploid cotton (Zhao et al. 2021). In allooctoploid strawberry, BSA-seq was used to specify the subgenomes origins of three male sterility QTL (Wada et al. 2021).

For autopolyploid species, a polyploid BSA-seq method was developed and tested using data from tetraploid potato and hexaploid sweetpotato. The minimum sequencing depth for identifying parent-specific simplex SNP calling was determined to be 40 × and 75 × for potato and sweetpotato, respectively (Yamakawa et al. 2021). Sequences from one parent are aligned to the species public genome to identify a reference SNP allele. Reads from the bulks and the second parent are reported in relation to reference alleles and SNP indices are calculated.

SNP loci were evaluated for both parents considering sites where SNP-indexes were equal to 0 in one of the parents, i.e., in nulliplex cases. The potato's *H1* resistance locus was identified on chromosome 5, and sweetpotato's anthocyanin QTL was detected on chromosome 12 (Yamakawa et al. 2021).

In our illustration, we have simply combined the read counts of BT progenies whose flesh color scores were lower than 4.30 (24 individuals within 'low' bulk) or greater than 7.21 (23 individuals within 'high' bulk) as if they were sequenced in their respective bulks out of the raw variant call format (VCF) files. From a total of 87,134 variants derived from *I. trifida* genome alignment, there were 8,567 and 18,226 SNPs in simplex (*Aaaaaa*) states for either 'Beauregard' or 'Tanzania', respectively, to contrast with the nulliplex (*aaaaaa*) states of the other parent. Differences between SNP-index from the low and high value bulks, ΔSNP-index, allowed us to detect the same regions on chromosome 3 and 12 to be associated with flesh color (Fig. 6.6).

6.4 Future Prospects

The main goal of QTL mapping is to investigate the genetic architecture of quantitative traits of interest. Single- and multiple-QTL models have been available for inbred, diploid mapping populations for quite some time now (see Da Costa and Zeng (2010) for a comprehensive review). However, only recently these methods became available for outbred, polyploid mapping populations. For sweetpotato, QTL mapping has been used in different populations (backgrounds) and for a set of traits to date. Great progress has been achieved, particularly with the recent studies which used molecular markers and statistical methods specifically developed for complex autopolyploid species. Next steps in QTL mapping should remain in the extension of the methods to account for multiple traits or environments simultaneously, enabling the investigation of pleiotropic effects and linkage as well as the interaction between QTL and environments. Certainly, the future results of these approaches will be helpful to improve

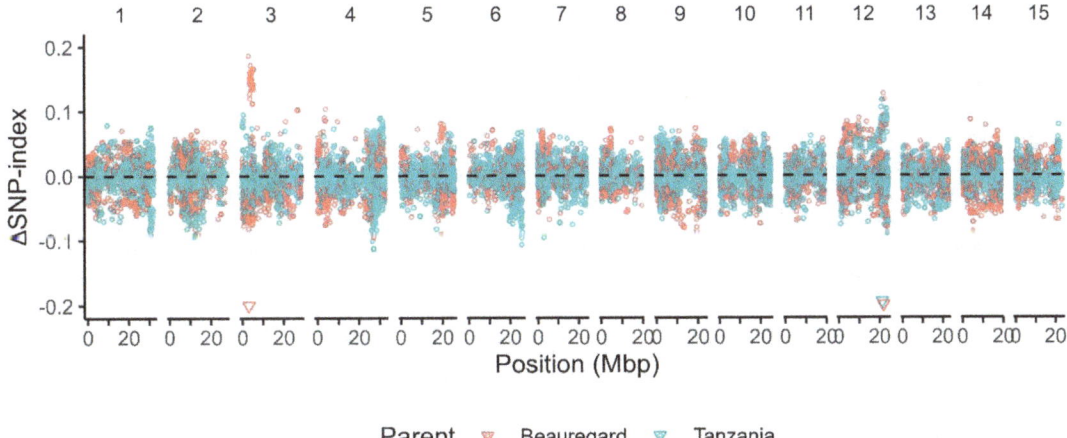

Fig. 6.6 BSA-seq for flesh color in sweetpotato 'Beauregard' × 'Tanzania' full-sib population (*N* = 298) evaluated in Peru. The evaluation of 26,793 simplex SNPs (8567 for 'Beauregard' and 18,226 for 'Tanzania') has shown highest absolute values of ΔSNP-index between 'low' and 'high' bulks on chromosome 3 (3,120,245 bp) and 12 (21,812,147 bp for 'Beauregard', and 21,271,480 bp for 'Tanzania') close to previously identified QTL regions

our understanding regarding genetic architecture of the traits. Major effects QTL, detected with stability in multiple environments and for several traits, could be incorporated as a fixed term into prediction statistical models, in the context of genomic selection. The development of these models is currently in progress for sweetpotato populations, in its first version, and the use of novel QTL could be helpful. Overall, we believe these approaches will provide valuable information for MAS into sweetpotato breeding.

References

Basten C, Weir B, Zeng Z-B, LT (1999) QTL cartographer version 1.13

Bourke PM, van Geest G, Voorrips RE, Jansen J, Kranenburg T, Shahin A, Visser RGF, Arens P, Smulders MJM, Maliepaard C (2018) PolymapR—linkage analysis and genetic map construction from F$_1$ populations of outcrossing polyploids. Bioinformatics 34(20):3496–3502. https://doi.org/10.1093/bioinformatics/bty371

Cervantes-Flores JC, Sosinski B, Pecota KV, Mwanga ROM, Catignani GL, Truong VD, Watkins RH, Ulmer MR, Yencho GC (2011) Identification of quantitative trait loci for dry-matter, starch, and β-carotene content in sweetpotato. Mol Breed 28:201–216. https://doi.org/10.1007/s11032-010-9474-5

Cervantes-Flores JC, Yencho GC, Kriegner A, Pecota KV, Faulk MA, Mwanga ROM, Sosinski BR (2008a) Development of a genetic linkage map and identification of homologous linkage groups in sweetpotato using multiple-dose AFLP markers. Mol Breed 21(4):511–532. https://doi.org/10.1007/s11032-007-9150-6

Cervantes-Flores JC, Yencho GC, Pecota KV, Sosinski B, Mwanga ROM (2008b) Detection of quantitative trait loci and inheritance of root-knot nematode resistance in sweetpotato. J Am Soc Hortic Sci 133(6):844–851. https://doi.org/10.21273/JASHS.133.6.844

Chang K-Y, Lo H-F, Lai Y-C, Yao P-J, Lin K-H, Hwang S-Y (2009) Identification of quantitative trait loci associated with yield-related traits in sweet potato (*Ipomoea batatas*). Bot Stud 50:43–55

Clevenger J, Chu Y, Chavarro C, Botton S, Culbreath A, Isleib TG, Holbrook CC, Ozias-Akins P (2018) Mapping late leaf spot resistance in peanut (*Arachis hypogaea*) using QTL-seq reveals markers for marker-assisted selection. Front Plant Sci 9:83. https://doi.org/10.3389/fpls.2018.00083

Collard BCY, Mackill DJ (2007) Marker-assisted selection: an approach for precision plant breeding in the twenty-first century. Philos Trans R Soc B Biol Sci 363(1491):557–572. https://doi.org/10.1098/rstb.2007.2170

Covarrubias-Pazaran G (2016) Genome-assisted prediction of quantitative traits using the R package sommer. PLoS ONE 11(6):e0156744. https://doi.org/10.1371/journal.pone.0156744

Da Costa E, Silva L, Zeng Z-B (2010) Current progress on statistical methods for mapping quantitative trait loci from inbred line crosses. J Biopharm Stat 20(2):454–481. https://doi.org/10.1080/10543400903572845

Da Silva Pereira G, Gemenet DC, Mollinari M, Olukolu BA, Wood JC, Diaz F, Mosquera V, Gruneberg WJ, Khan A, Buell CR, Yencho GC, Zeng Z-B (2020) Multiple QTL mapping in autopolyploids: A random-effect model approach with application in a hexaploid sweetpotato full-sib population. Genetics 215(3):579–595. https://doi.org/10.1534/genetics.120.303080

Da Silva Pereira G, Oloka BM, Ssali R, Yada B, Dos Santos IG, Rosa JRBF, Silva CC, Dias KO das G, Azevedo CF, Olukolu BA, De Boeck B, Campos H, Yencho C (2023) Use of advanced analytics and genomic selection for sweetpotato breeding. In: 30 Plant and Animal Genome Conference. San Diego

Doerge RW, Churchill GA (1996) Permutation tests for multiple loci affecting a quantitative character. Genetics 142(1):285–294. https://doi.org/10.1093/genetics/142.1.285

Edwards MD, Stuber CW, Wendel JF (1987) Molecular-marker-facilitated investigations of quantitative-trait loci in maize. I. Numbers, genomic distribution and types of gene action. Genetics 116(1):113–125. https://doi.org/10.1093/genetics/116.1.113

Ehrenreich IM, Torabi N, Jia Y, Kent J, Martis S, Shapiro JA, Gresham D, Caudy AA, Kruglyak L (2010) Dissection of genetically complex traits with extremely large pools of yeast segregants. Nature 464(7291):1039–1042. https://doi.org/10.1038/nature08923

Gemenet DC, da Silva Pereira G, De Boeck B, Wood JC, Mollinari M, Olukolu BA, Diaz F, Mosquera V, Ssali RT, David M, Kitavi MN, Burgos G, Felde TZ, Ghislain M, Carey E, Swanckaert J, Coin LJM, Fei Z, Hamilton JP, Yada B, Yencho GC, Zeng Z-B, Mwanga ROM, Khan A, Gruneberg WJ, Buell CR (2020a) Quantitative trait loci and differential gene expression analyses reveal the genetic basis for negatively associated β-carotene and starch content in hexaploid sweetpotato [*Ipomoea batatas* (L.) Lam.]. Theor Appl Genet 133(1):23–36. https://doi.org/10.1007/s00122-019-03437-7

Gemenet DC, Lindqvist-Kreuze H, De Boeck B, da Silva Pereira G, Mollinari M, Zeng Z-B, Craig Yencho G, Campos H (2020b) Sequencing depth and genotype quality: accuracy and breeding operation considerations for genomic selection applications in autopolyploid crops. Theor Appl Genet 133(12):3345–3363. https://doi.org/10.1007/s00122-020-03673-2

Grattapaglia D, Sederoff R (1994) Genetic linkage maps of *Eucalyptus grandis* and *Eucalyptus urophylla* using a pseudo-testcross: mapping strategy and

RAPD markers. Genetics 137(4):1121–1137. https://doi.org/10.1093/genetics/137.4.1121

Grüneberg WJ, Eyzaguirre R, Díaz F, Boeck B de, Espinoza J, Swanckaert J, Dapaah H, Andrade MI, Makunde GS, Agili S, Ndingo-Chipungu FP, Attaluri S, Kapinga R, Nguyen T, Kaiyung X, Tjintokohadi K, Ssali RT, Carey T, Low JW, Mwanga ROM (2019) Procedures for the evaluation of sweetpotato trials. International Potato Center

Hackett CA, Bradshaw JE, Bryan GJ (2014) QTL mapping in autotetraploids using SNP dosage information. Theor Appl Genet 127(9):1885–1904. https://doi.org/10.1007/s00122-014-2347-2

Jansen RC, Stam P (1994) High resolution of quantitative traits into multiple loci via interval mapping. Genetics 136(4):1447–1455. https://doi.org/10.1093/genetics/136.4.1447

Jiang C, Zeng Z-B (1997) Mapping quantitative trait loci with dominant and missing markers in various crosses from two inbred lines. Genetica 101(1):47–58. https://doi.org/10.1023/A:1018394410659

Kao C-H, Zeng Z-B, Teasdale RD (1999) Multiple interval mapping for quantitative trait loci. Genetics 152(3):1203–1216. https://doi.org/10.1093/genetics/152.3.1203

Lander ES, Botstein D (1989) Mapping mendelian factors underlying quantitative traits using RFLP linkage maps. Genetics 121(1):185–199. https://doi.org/10.1093/genetics/121.1.185

Li H, Zhao N, Yu X, Liu Y, Zhai H, He S, Li Q, Ma D, Liu Q (2014) Identification of QTLs for storage root yield in sweetpotato. Sci Hortic 170:182–188. https://doi.org/10.1016/j.scienta.2014.03.021

Lindqvist-Kreuze H, Bonierbale M, Grüneberg WJ, Mendes T, De Boeck B, Campos H (2023) Potato and sweetpotato breeding at the International Potato Center: approaches, outcomes and the way forward. Theor Appl Genet 137(1):12. https://doi.org/10.1007/s00122-023-04515-7

Michelmore RW, Paran I, Kesseli RV (1991) Identification of markers linked to disease-resistance genes by bulked segregant analysis: a rapid method to detect markers in specific genomic regions by using segregating populations. Proc Natl Acad Sci 88(21):9828–9832. https://doi.org/10.1073/pnas.88.21.9828

Mollinari M, Garcia AAF (2019) Linkage analysis and haplotype phasing in experimental autopolyploid populations with high ploidy level using hidden markov models. G3 Genes Genomes Genetics 9(10):3297–3314. https://doi.org/10.1534/g3.119.400378

Mollinari M, Olukolu BA, da Silva Pereira G, Khan A, Gemenet D, Yencho GC, Zeng Z-B (2020) Unraveling the hexaploid sweetpotato inheritance using ultra-dense multilocus mapping. G3 Genes Genomes Genet 10(1):281–292. https://doi.org/10.1534/g3.119.400620

Mwanga ROM, Odongo B, p'Obwoya CO, Gibson RW, Smit NEJM, Carey EE (2001) Release of five sweetpotato cultivars in Uganda. HortScience 36(2):385–386. https://doi.org/10.21273/HORTSCI.36.2.385

Mwanga ROM, Swanckaert J, da Silva Pereira G, Andrade MI, Makunde G, Grüneberg WJ, Kreuze J, David M, De Boeck B, Carey E, Ssali RT, Utoblo O, Gemenet D, Anyanga MO, Yada B, Chelangat DM, Oloka B, Mtunda K, Chiona M, Koussao S, Laurie S, Campos H, Yencho GC, Low JW (2021) Breeding progress for vitamin A, iron and zinc biofortification, drought tolerance, and sweetpotato virus disease resistance in sweetpotato. Front Sustain Food Syst 5:616674. https://doi.org/10.3389/fsufs.2021.616674

Oloka BM, da Silva Pereira G, Amankwaah VA, Mollinari M, Pecota KV, Yada B, Olukolu BA, Zeng Z-B, Craig Yencho G (2021) Discovery of a major QTL for root-knot nematode (*Meloidogyne incognita*) resistance in cultivated sweetpotato (*Ipomoea batatas*). Theor Appl Genet 134(7):1945–1955. https://doi.org/10.1007/s00122-021-03797-z

Park M, Lee J-H, Han K, Jang S, Han J, Lim J-H, Jung J-W, Kang B-C (2019) A major QTL and candidate genes for capsaicinoid biosynthesis in the pericarp of Capsicum chinense revealed using QTL-seq and RNA-seq. Theor Appl Genet 132(2):515–529. https://doi.org/10.1007/s00122-018-3238-8

Qu L, Guennel T, Marshall SL (2013) Linear score tests for variance components in linear mixed models and applications to genetic association studies. Biometrics 69(4):883–892. https://doi.org/10.1111/biom.12095

Rolston LH, Clark CA, Cannon JM, Randle WM, Riley EG, Wilson PW, Robbins (1987) Beauregard' sweet potato. HortScience 22(6):1338–1339. https://doi.org/10.21273/HORTSCI.22.6.1338

Rosyara UR, De Jong WS, Douches DS, Endelman JB (2016) Software for genome-wide association studies in autopolyploids and Its application to potato. Plant Genome 9(2). https://doi.org/10.3835/plantgenome2015.08.0073

Stuber CW, Edwards MD, Wendel JF (1987) Molecular marker-facilitated investigations of quantitative trait loci in maize. II. Factors influencing yield and its component traits. Crop Sci 27(4). https://doi.org/10.2135/cropsci1987.0011183X002700040006x

Takagi H, Abe A, Yoshida K, Kosugi S, Natsume S, Mitsuoka C, Uemura A, Utsushi H, Tamiru M, Takuno S, Innan H, Cano LM, Kamoun S, Terauchi R (2013) QTL-seq: rapid mapping of quantitative trait loci in rice by whole genome resequencing of DNA from two bulked populations. Plant J 74(1):174–183. https://doi.org/10.1111/tpj.12105

van Ooien JW, Boer MP, Jansen R, Maliepaard C (2000) MapQTL 4.0: software for the calculation Of QTL positions on genetic maps. Plant Research International, Wageningen

Voorrips RE, Gort G, Vosman B (2011) Genotype calling in tetraploid species from bi-allelic marker data using mixture models. BMC Bioinform 12(1):172. https://doi.org/10.1186/1471-2105-12-172

Wada T, Monden H, Isobe S, Shirasawa K, Sueyoshi T, Hirata C, Mori M, Nagamatsu S, Tanaka Y (2021) Comparative QTL mapping for male sterility of cultivated strawberry (*Fragaria* × *ananassa* Duch.) using different reference genome sequences. Breed Sci 71(4):456–466. https://doi.org/10.1270/jsbbs.20151

Wang Z, Yan L, Chen Y, Wang X, Huai D, Kang Y, Jiang H, Liu K, Lei Y, Liao B (2022) Detection of a major QTL and development of KASP markers for seed weight by combining QTL-seq, QTL-mapping and RNA-seq in peanut. Theor Appl Genet 135(5):1779–1795. https://doi.org/10.1007/s00122-022-04069-0

Wen J, Jiang F, Weng Y, Sun M, Shi X, Zhou Y, Yu L, Wu Z (2019) Identification of heat-tolerance QTLs and high-temperature stress-responsive genes through conventional QTL mapping, QTL-seq and RNA-seq in tomato. BMC Plant Biol 19(1):398. https://doi.org/10.1186/s12870-019-2008-3

Wu S, Lau KH, Cao Q, Hamilton JP, Sun H, Zhou C, Eserman L, Gemenet DC, Olukolu BA, Wang H, Crisovan E, Godden GT, Jiao C, Wang X, Kitavi M, Manrique-Carpintero N, Vaillancourt B, Wiegert-Rininger K, Yang X, Bao K, Schaff J, Kreuze J, Gruneberg W, Khan A, Ghislain M, Ma D, Jiang J, Mwanga ROM, Leebens-Mack J, Coin LJM, Yencho GC, Buell CR, Fei Z (2018) Genome sequences of two diploid wild relatives of cultivated sweetpotato reveal targets for genetic improvement. Nat Commun 9(1):4580. https://doi.org/10.1038/s41467-018-06983-8

Yada B, Brown-Guedira G, Alajo A, Ssemakula GN, Owusu-Mensah E, Carey EE, Mwanga RO, Yencho GC (2017) Genetic analysis and association of simple sequence repeat markers with storage root yield, dry matter, starch and β-carotene content in sweetpotato. Breed Sci 67(2):140–150. https://doi.org/10.1270/jsbbs.16089

Yamakawa H, Haque E, Tanaka M, Takagi H, Asano K, Shimosaka E, Akai K, Okamoto S, Katayama K, Tamiya S (2021) Polyploid QTL-seq towards rapid development of tightly linked DNA markers for potato and sweetpotato breeding through whole-genome resequencing. Plant Biotechnol J 19(10):2040–2051. https://doi.org/10.1111/pbi.13633

Yan H, Ma M, Ahmad MQ, Arisha MH (2022) High-density single nucleotide polymorphisms genetic map construction and quantitative trait locus mapping of color-related traits of purple sweet potato [*Ipomoea batatas* (L.) Lam.]. Front Plant Sci 12. https://doi.org/10.3389/fpls.2021.797041

Yu X, Zhao N, Li H, Jie Q, Zhai H, He S, Li Q, Liu Q (2014) Identification of QTLs for starch content in sweetpotato (*Ipomoea batatas* (L.) Lam.). J Integr Agric 13(2):310–315. https://doi.org/10.1016/S2095-3119(13)60357-3

Zeng ZB (1994) Precision mapping of quantitative trait loci. Genetics 136(4):1457–1468. https://doi.org/10.1093/genetics/136.4.1457

Zhang X, Wang W, Guo N, Zhang Y, Bu Y, Zhao J, Xing H (2018) Combining QTL-seq and linkage mapping to fine map a wild soybean allele characteristic of greater plant height. BMC Genomics 19(1):226. https://doi.org/10.1186/s12864-018-4582-4

Zhao N, Yu X, Jie Q, Li H, Li H, Hu J, Zhai H, He S, Liu Q (2013) A genetic linkage map based on AFLP and SSR markers and mapping of QTL for dry-matter content in sweetpotato. Mol Breed 32(4):807–820. https://doi.org/10.1007/s11032-013-9908-y

Zhao Y, Chen W, Cui Y, Sang X, Lu J, Jing H, Wang W, Zhao P, Wang H (2021) Detection of candidate genes and development of KASP markers for *Verticillium wilt* resistance by combining genome-wide association study, QTL-seq and transcriptome sequencing in cotton. Theor Appl Genet 134(4):1063–1081. https://doi.org/10.1007/s00122-020-03752-4

Zou F, Fine JP, Hu J, Lin DY (2004) An efficient resampling method for assessing genome-wide statistical significance in mapping quantitative trait loci. Genetics 168(4):2307–2316. https://doi.org/10.1534/genetics.104.031427

Genomic-Assisted Nutritional and Quality Breeding Efforts in Sweetpotato

7

Reuben Tendo Ssali, Bonny Michael Oloka, Victor A. Amankwaah, Benard Yada, Edward Carey, and G. Craig Yencho

Abstract

Sweetpotato, *Ipomoea Batatas* L., is widely considered as a staple food in many developing countries especially in sub-Saharan Africa (SSA). It has storage roots that are rich in starch, micronutrients (especially Fe and Zn), and vitamins A and C. Partly due to its polyploidy, heterogeneity and outcrossing nature, it can be cultivated in a wide range of environments. Sweetpotato additionally comes in a variety textures, colors, flavors, and nutritional profiles thus appealing to many different groups of people. Conventional breeding of sweetpotato to improve nutritional and quality traits has only been moderately successful over the years due to several factors. This is manly true in areas like SSA which are characterized by low adoption of high β-carotene orange-flesh types. The biggest challenge has been to understand, measure and combine cultural preferences for taste and quality with nutritional components. In this work, we look at efforts to measure quality preferences, and accelerate the breeding process of sweetpotato by tapping into new breeding technologies and genomic-assisted breeding approaches.

R. T. Ssali (✉)
International Potato Center (CIP), Kampala, Uganda
e-mail: R.Ssali@cgiar.org

B. M. Oloka · G. C. Yencho
Department of Horticultural Science, North Carolina State University, Raleigh, NC 27695, USA
e-mail: boloka@ncsu.edu

G. C. Yencho
e-mail: craig_yencho@ncsu.edu

V. A. Amankwaah
CSIR-Crops Research Institute, Kumasi, Ghana
e-mail: va.amankwaah@gmail.com

B. Yada
National Crops Resources Research Institute, National Agricultural Research Organization, Kampala, Uganda
e-mail: yadabenard21@gmail.com

E. Carey
International Potato Center (CIP), Kumasi, Ghana
e-mail: tcarey@ksu.edu

Keywords

Sweetpotato nutrition · Cooking quality · Sensory evaluation · Textural characteristics

7.1 Introduction

Sweetpotato (*I. batatas* L.) is a widely grown staple crop in the tropical and subtropical developing world. It ranks seventh in global food production and fifth in production within developing countries. In 2022, world production of sweetpotato was 86.4 million metric tons, from a total harvested area of 7.4 million hectares

G. C. Yencho et al. (eds.), *The Sweetpotato Genome*, Compendium of Plant Genomes,
https://doi.org/10.1007/978-3-031-65003-1_7

(FAOSTAT 2022). The developing countries account for 95% of this production. Sweetpotato is often considered a crop for low-income farmers because it is relatively easy to produce with minimal inputs and is known for its ability to produce high yields in marginal environments. It is grown in diverse agroecological areas from the desert edges of lowland tropics to the highlands of the humid tropics. This adaptability to diverse environmental conditions is due to the inherent plasticity of the crop and the range of genotypes grown. The crop is ideal for subsistence farmers because maintaining planting material is straight forward. Although sweetpotato is primarily used as a source of carbohydrates from the tuberous roots, there is growing awareness of the additional health benefits that come from the consumption of these storage roots. It is considered as a staple food in many developing countries due to its starch-rich storage roots (da Silva Pereira et al. 2023).

Generally, the roots have a high moisture content with an average dry matter content of 25–30% (Truong et al. 2018). Starch is the major carbohydrate in sweetpotato storage roots, making up approximately 80–90% of the dry matter (Tumwegamire et al. 2011a, b). Sugars make up about 15–20% of the dry matter and are mostly in the form of sucrose, glucose, fructose, and maltose, with the latter being undetectable in raw roots but predominant in cooked roots due to β-amylolysis of starch (Kitahara et al. 2017; Truong et al. 2018; Amankwaah 2019; Amankwaah et al. 2023). It is also rich in mineral content with potassium, manganese, copper, iron and zinc. Potassium is the most abundant mineral, available at concentrations as high as 300 mg/100 g of fresh roots. This can contribute about a fifth of the recommended dietary allowance for children (Sanoussi et al. 2016). The mineral content is not only dependent on the variety and cooking conditions but also on agricultural practices, particularly the use of fertilizers. Sweetpotato also contains high-quality proteins (Truong et al. 2018), rich in methionine, threonine, valine,

and tryptophan (de Albuquerque et al. 2019). The protein content is low in comparison with cereal crops (Neela and Fanta 2019). It has been observed that baking sweetpotato roots reduces their protein content whereas boiling helps retain it. Sweetpotato provides twice the amount of dietary fiber of potato and cassava (Neela and Fanta 2019) thus aiding in satiety and digestive health. Some genotypes are rich in pro-vitamin A (β-carotene) and vitamins B1, B3, C, and E (Woolfe 1992; Truong et al. 2018; Neela and Fanta 2019). Therefore, sweetpotato can be exploited through targeted breeding to increase nutrient content for the world's food needs (Mwanga et al. 2021a, b). This nutrient density underscores its importance as a staple food and an ally in addressing hunger, malnutrition, and poverty in sub-Saharan Africa. Specifically, sweetpotato diversifies diets, provides sustenance to vulnerable populations, and offers income opportunities for smallholder farmers.

The nutritional value of a food crop depends largely on its beneficial nutrients, organoleptic properties, and any undesirable elements. Conversely, nutritional quality is determined by its impact on the nutritional status and health of those who consume it. Therefore, a crop can be nutritionally valuable but not necessarily of high nutritional quality if it's not accessible or preferred by the community. This distinction is crucial for enhancing sweetpotato varieties. Sweetpotato varieties with high nutritional value that are easy to grow and more affordable offer significant benefits. However, to boost the nutritional quality of sweetpotato, breeding efforts should also prioritize traits like taste, ease of preparation, and yield. It's important to note that high nutritional value and quality can coexist, and improving both could positively affect the health and livelihoods of subsistence farmers who rely on sweetpotato. Nonetheless, it is essential to identify the nutritional traits that need enhancement and to understand and quantify the genetic factors influencing these traits. Historically, these aspects have not been the primary focus in the sweetpotato variety development.

Although sweetpotato is an important food crop in the tropics and sub-tropics, it often faces significant challenges from pests and diseases. This affects its production and availability of planting materials, especially for high-yielding and nutritionally rich varieties. Breeding for pest and disease resistance along with improved yield and nutritional quality, is essential for enhancing the livelihoods of sweetpotato farmers and consumers. The success of new sweetpotato varieties relies on their production characteristics, but most importantly on their sensory and utilization qualities for the consumers (Tomlins et al. 2007). Consequently, prioritizing the preferred characteristics of sweetpotato roots for the end users is a key objective in sweetpotato breeding. Furthermore, the sweetpotato food chain is characterized by different actors with different preferences of choice due to factors such as socio-economic and gender dynamics (Mudege et al. 2019). Therefore, it is crucial to define preferences for quality characteristics by market segments and gender to meet the diverse needs of end users. For instance, Ugandan consumers of boiled and/or steamed sweetpotato prefer varieties that are aromatic, sweet, mealy, firm, and non-fibrous (Mwanga et al. 2021a, b). However, these desirable qualities must be linked to the biophysical and functional properties of the food to develop laboratory methods for quantitative evaluation. Typically, sweetpotato breeding programs assess these nutritional, physical, dietary and cooking attributes only in the final stages of diversity testing and release when 90–99% of breeding lines have already been discarded (Kays and Wang 2002; Grüneberg et al. 2015). This is due to the cost and complexity of these analyses and the lack of high-throughput phenotyping protocols and equipment (Kays and Wang 2002). Therefore developing high-throughput phenotyping protocols for user-approved quality and nutritional traits will facilitate their integration into breeding and selection of sweetpotato.

Genomic-assisted breeding is a promising approach for developing sweetpotato varieties with improved nutritional, eating, and processing qualities right from the early breeding stages. This approach involves using genomic information to identify genetic markers associated with desirable traits. The chapter describes progress towards the application of genomic assisted breeding to develop new varieties with improved eating quality, processing suitability and nutrition value of sweetpotato. In contrast to conventional breeding, which depends on the visual assessment of traits to choose preferred characteristics, genomic-assisted breeding employs a more precise strategy.

7.2 Importance of Nutritional Value in Sweetpotato

Billions of people in developing countries suffer from chronic deficiencies in essential nutrients. Scientists have been working towards designing crops with improved essential nutrient content through either agricultural practices also known as "agronomic biofortification" or breeding (Bouis and Welch 2010). Micronutrient deficiencies are widespread, and diet-based strategies are the most sustainable solutions. These deficiencies affect particularly significant proportions of populations in developing countries and are especially high in pregnant women and pre-school children. In the developing world, an estimated 122 countries have populations deficient in Vitamin A. Sub-Saharan Africa has the highest percentage, with 48% of children under 5 affected. Each year, between 250,000 to 500,000 malnourished children go, with about half of them dying within a year of losing their sight. In 24 countries across Africa and Southeast Asia, over 20% of pregnant women suffer from night blindness and severe vitamin A deficiency is a major cause of maternal mortality in these regions. Incorporating sweetpotato into the diet is a practical solution for vitamin A deficiency. Since sweetpotato is commonly grown in the developing world, introducing pro-Vitamin A rich varieties would be an effective and low-cost strategy to combat vitamin A deficiency. All plants have high levels of carotenoids in their leaves, primarily in thylakoid membranes. Sweetpotato leaves specifically contain four carotenoids: lutein (47.6% of total carotenoids), β-carotene (25.2%), violaxanthin (13.9%) and neoxanthin (9.6%) (Chen and Chen 1993). Khan et al. (2022) suggested that use of molecular genetic and genomic methods to understand and manipulate carotenoid biosynthesis in sweetpotato might provide the fastest and most effective means to generate increased pro-vitamin A sweetpotato cultivars.

Biofortification is a health-based strategy that aims to reduce micronutrient deficiencies and improve public health through the development of staple food crops that are rich in essential vitamins and minerals. The approach is to either enrich or increase the accumulation of micronutrients in the storage organs of the crop through plant breeding, using traditional methods or genetic engineering. Biofortification is generally perceived as a more sustainable, cost-effective, and efficient means to alleviate malnutrition than providing nutrient supplements or commercially fortified foods (Bouis and Welch 2010; Haas et al. 2005) and molecular genetics and genomic information presents new opportunities to expedite biofortification of crops (Bouis and Saltzman 2017; Welch and Graham 2004).

Although sweetpotato is primarily grown for its storage root, the young leaves and shoots are also consumed in many countries in sub-Saharan Africa, the Pacific, Asia, and the Caribbean. Sweetpotato is widely regarded as a healthy food and ranks high in nutritional value among root and tuber crops and as a staple food. It is a rich source of carbohydrates, dietary fiber, vitamins A, C, and several B-vitamins. The roots are fat-free and contain a moderate source of complex starch, which makes it an ideal food for diabetics. However, the most significant nutritional attribute of sweetpotato is the high beta-carotene content found in orange-fleshed varieties, which is a dietary source of vitamin A.

7.3 Genetic Basis for Nutritional Value in Sweetpotato

A crucial first step to conduct biofortification of crops is to gain a better understanding of the genetic and physiological factors affecting variation in nutrient content. Until relatively recently most plant breeders have employed a "phenotypic" approach to enhance nutrient content, which essentially involves identifying plants with desirable nutrient content based on visual observation of proxy traits like flesh color or laboratory analysis and using these plants as a source of genes for nutrient enhancement. This approach is limited by low heritabilities of many nutritional traits, difficulties in accurate and rapid phenotypic assessment of nutrient content, and problems with obtaining desirable levels of nutrient content in combination with other traits of agronomic importance (Bouis and Saltzman 2017).

The exploration of the genetic underpinnings that contribute to the nutritional value of sweetpotato is both intricate and intriguing. Sweetpotato is characterized by its autohexaploid nature, possessing 90 somatic chromosomes ($2n = 6x = 90$), and a substantial genome size estimated at approximately 4.4 Gb. The genetic structure of sweetpotato is a determinant of its nutritional profile, which includes mineral content (like Fe and Zn), vitamins (such as A and C), and nutraceutical components such as β-carotene and anthocyanins, especially notable in varieties with orange and purple flesh (da Silva Pereira et al. 2023). These elements are vital for human well-being and contribute to the agricultural value of the crop. The recent progress in developing molecular tools has enabled genetic mapping of significant quantitative trait loci (QTLs) and a genome-wide characterization of population structure, paving the way for genomics-assisted breeding. This approach is designed to develop new varieties that not only yield abundantly in adverse conditions but also exhibit resistance to pests and diseases, alongside enhanced nutritional properties.

The great biological variation in the genus Ipomoea, its complex genetic structure, and its high ploidy levels make the study of the inheritance of specific traits and the gene mapping difficult. Although great improvement was achieved in the last decade for studying and understanding the molecular composition of the sweetpotato genome (analysis of DNA content, construction of genetic maps, and development of genomic and expressed sequenced tags resources), the complex inheritance of the nutritional traits in sweetpotato (vitamin A, vitamin C, E, dietary fiber, glycemic index, and anthocyanin content) demands new strategies for dissecting them. Whatever the trait, due to the complexity of the root's genotype and the environmental interactions, the understanding of the regulatory or structural genes and the allelic polymorphisms that control them is not enough for the development of a functional marker and an effective regulatory system for improvement, but it is certainly an essential step in the right direction. At this point, a thorough understanding of the gene functions and biochemical pathways by which the nutrient synthesis and accumulation are controlled is lacking for sweetpotato. Such information is a prerequisite for devising strategies to increase the nutritional quality. Since each nutritional trait can be affected by many different factors, it is not feasible to discuss all of them in this context. A more general view of where we stand and where we should go for the better understanding of the genetic regulation of sweetpotato nutritional traits would be explained in the following paragraphs. Despite the great nutritional potential of sweetpotato, its genetic background and the regulatory mechanisms that control the nutritional trait composition are still poorly understood. The recent accomplishment of gene interrogations and functional genomics approaches in the model plants and the development of the high-throughput methods including the transcriptomics and metabolomics has heightened the momentum to study the nutritional traits of sweetpotato at the molecular level. The information gleaned will aid in the breeding approaches mentioned in the other papers of this publication and be a platform to describe the desirable nutrient-dense sweetpotato of the future.

As the nutritional value of sweetpotato is directly related to human consumption and directly influences the livelihood of developing nations, there has been interest in determining the genetic markers to the nutritional value. Through identification of such markers, breeders could select for higher nutritional value sweetpotato, and farmers could grow sweetpotato specifically suited for certain nutritional needs. However, determining the nutritional value of storage roots is time consuming and expensive because chemical analyses must be run on large sample sizes to produce accurate results. There have been efforts of using near-infrared reflectance spectroscopy (NIRS) to predict storage root nutritional value because it is a rapid, nondestructive, inexpensive method of analysis and NIRS has been successful in predicting various traits in other crops. Ash of the storage root, which is the mineral content, has been the most successful in terms of accurate and precise prediction of nutritional value using NIRS compared to other nutritional factors. It was demonstrated by Tumwegamire et al. (2011a; b) that NIRS could be used to reliably predict dry matter, starch, and beta-carotene content in sweetpotato storage roots. Ash prediction equations could then be used to see if there are quantitative trait loci (QTL) affecting this nutritional factor. Development of genetic markers for nutritional value will allow breeders to breed for high nutritional value sweetpotatoes more efficiently.

As mentioned earlier, it is very difficult to define a superior sweetpotato genotype for nutritional value since high nutritional value is largely dependent on having a good balance between various nutritional components. However, it is possible to breed sweetpotato to have a specific nutritional composition if this is desired. It is basically a matter of setting levels for specific nutritional parameters and then selecting progeny that meet the desired nutritional profile. This is easily achieved for Pro-vitamin A since various flesh colors can be attributed with specific carotenoid content levels; hence, the only requirement is to select progeny that have a specific flesh color. However, selecting progeny based on sensory data to ensure acceptability of nutritious varieties is problematic since sensory data will always have a large environmental effect but an understanding of which genes are affected by the environment will enable the development of genotypes that have a specific nutritional profile under defined storage conditions. Since it is genes that control nutritional composition, determining specific genes for each nutritional parameter will be the ultimate way to modify sweetpotato nutritional composition. This can be done using a QTL analysis or by using a transgenic approach. While transgenics will give an instant result, the development of a transgenic approach is not feasible for sweetpotato breeding in developing countries, and a transgenic approach also has social and ethical issues. A QTL analysis, while being a long process, will enable the development of genotypes with specific nutritional profiles without many of the negative issues of transgenics. So both of these methods have their advantages, and it is fortuitous that both can be employed to determine the specific genes for each nutritional parameter, with the knowledge of a QTL analysis being transferable to a marker-assisted selection (MAS) program.

A number of key genes responsible for different nutritional components of sweetpotato have been identified, including (1) **Su α-branching enzyme gene** responsible for amylose content (Takiko et al. 2006); **sporamin gene** involving storage root formation and it is rich in methionine content (Ravi et al. 2014); **β-amylase gene** controlling sugar and starch content in the roots (Nakamura et al. 2014); **IbVIN1** encoding a vacuolar invertase that limits sucrose accumulation in the roots at low temperature (Ru et al. 2021). The knowledge of these genes enables better understanding of effect on nutritional traits, provides directions for selection or breeding of sweetpotato with improved nutritional quality.

It is recognized that the nutritional quality and health benefit of any crop will depend on the various bioactive constituents and antinutritional factors, as well as the inherent diversity in human nutritive and health needs. A detailed understanding of the genes encoding the enzymes and other proteins that dictate the nutritional and health status of a crop is a prerequisite for crop improvement. The successful development of nutritionally improved crops will require a comprehensive understanding of the genetic control of the target nutritional traits. Over the past two decades, significant research efforts have been directed at understanding the genetic basis for nutritional quality in several crops. Although this research has been fragmented, a relatively coherent and detailed picture has emerged in some cases. This has largely been due to the advent and application of genomics and other bioinformatic tools. While the application of these tools to sweetpotato lags far behind other major crops, there have been some notable research initiatives on the genetic basis for nutritional quality in sweetpotato. In some cases, findings from research on other crops has been fortuitously useful to sw2020aeetpotato researchers. For example, knowledge of the genes encoding the enzymes of starch biosynthesis is fairly advanced in several major crops. This has allowed sweetpotato researchers, on occasion, to glean useful information relevant to sweetpotato without actually doing the research. A case in point is a study aimed at understanding the genetic control of starch quantity and quality in sweetpotato (Gemenet et al. 2020b). The aim of the study was to identify genes encoding enzyme of starch biosynthesis which are differentially expressed in storage root of a high versus low starch line. The study exploited the fact that sequence information and PCR primers for such genes were available from similar studies in other crops, notably maize and potato. Using these, the researchers were able to identify the sweetpotato counterpart genes and address the research aims. Future research efforts on the genetic basis for nutritional quality of sweetpotato would be greatly facilitated by access to information on genes and gene families of other crops, and the development of sweetpotato genomics and bioinformatics.

The final level of a biological system involved in the expression of phenotype, from both genetic and environmental impacts, is the accumulation of specific gene products and their interactions to form the unique attributes of a defined cell type or tissue. Gene expression and its regulation represent the process by which a gene's DNA sequence is transcribed into RNA and the RNA transcript is then translated into a protein. Changes in the dynamics of gene regulation, including changes in mRNA levels, and changes in protein function, can greatly influence phenotype. Changes in gene expression can have both direct and indirect effects on phenotype. Currently, gene regulation is assumed to represent the most frequent form of natural variation within a species, and the easiest means by which complex phenotypic variation can occur. This assumes importance as genotyping efforts begin to identify genes and QTL which control specific nutritional traits, particularly in centers of crop diversity where landraces may have high nutritional value but lack the needed consumer preferred traits. Ultimately, these genes or QTL must be identified in breeding material, and fully characterized, in order to exploit them for crop improvement.

7.4 Phenotyping Tools for Nutritional and Cooking Quality Traits

Phenotyping tools are essential for characterizing and measuring the nutritional and cooking quality traits of sweetpotato. Here are some commonly used phenotyping tools for these traits:

1. Dry Matter Content: Dry matter content refers to the amount of solid material in sweetpotato. This is an important trait for assessing cooking quality, as it can affect the texture and flavor of the cooked sweetpotato. Dry matter content can be measured using techniques like oven-drying or freeze-drying (Twegamire et al. 2011a; b; Gruneberg et al. 2019).
2. Starch Content: Starch content is an important determinant of cooking quality, as it

affects the texture and flavor of the cooked sweetpotato. Starch content can be measured using techniques like HPLC (High-Performance Liquid Chromatography) or a simple iodine test (Twegamire et al. 2011a; b).
3. Sugar Content: Sweetpotato are known for their sweetness, which is due to their high sugar content. Sugar content can be measured using techniques like HPLC or refractometry or NIRS (Twegamire et al. 2011a; b).
4. Antioxidant Content: Sweetpotato are rich in antioxidants, which can protect against chronic diseases like cancer and heart disease. Antioxidant activity can be measured using techniques like DPPH (2,2-diphenyl-1-picrylhydrazyl) assay or spectrophotometry (Shimamura et al. 2014).
5. Vitamin and Mineral Content: Sweetpotato are a rich source of vitamins and minerals like vitamin A, vitamin C, potassium, and iron. These can be measured using techniques like HPLC or spectrophotometry (Twegamire et al. 2011a; b).
6. Color: The color of sweetpotato can be an indicator of their nutritional value, with darker varieties containing more antioxidants. Color can be measured using techniques like colorimetry and image analysis (Nakatumba-Nabende et al. 2023).
7. Texture: Texture is an important determinant of cooking quality, as it affects the mouthfeel of the cooked sweetpotato. Texture can be measured using techniques like texture profile analysis or sensory evaluation or image analysis (Nakitto et al. 2022; Nakatumba-Nabende et al. 2023).

Sensory characteristics of sweetpotato roots are critical to consumer choice and acceptability with potential to drive the adoption of improved varieties (Jenkins et al. 2018; Mwanga et al. 2021a, b). Sensory evaluation has been defined by the Institute of Food Technologists as a scientific method used to evoke, measure, analyze, and interpret responses to products as perceived through the senses of sight, hearing, touch, smell, and taste (IFT 2007). It can either be objective or subjective; objective evaluation (also known as descriptive sensory evaluation) makes use of trained sensory panelists to rate eating quality differences whereas in subjective evaluation (hedonic) consumers' reactions to the sensory properties of products are measured (Kemp et al. 2018). Consumer testing provides invaluable information regarding potential acceptance or rejection, and the reasons for rejection by consumers. While descriptive sensory analysis gives detailed and reliable information about the intensity of the quality attributes of a product, thus providing a basis for understanding acceptability (Joanna et al. 2019).

Recently, a systematic deployment of descriptive sensory evaluation for sweetpotato breeding was described in Uganda (Nakitto et al. 2022). This involved (1) lexicon development, (2) panel training and (3) evaluation of genotypes. A lexicon was developed for sweetpotato comprising 27 sensory attributes for characterization and differentiation of genotypes by sensory profiles (Table 7.1).

7.5   Improving Bioavailability of Nutrients

Bioavailability is a measure of the degree and rate at which a nutrient is absorbed from the diet and used for normal body functions (Jackson 1997). It is an intricate concept with the possibility of multiple nutrients interacting within the body to influence the absorption of each other. The rate of metabolism can also influence the availability of a given nutrient. Transitory complexes of starch, which are not directly correlated to yield of a storage root, can provide a readily mobilized source of energy and increase the bioavailability of other nutrients by sparing them from being used as an energy source. The identification of key enzymes and transporters in synthesis and accumulation of the targeted nutrient can lead to transgenic approaches specifically aimed at increasing the nutritive value of the crop (Shahzad et al. 2021). An example from sweetpotato would be the conversion of a portion of the storage root anthocyanin pigments into proanthocyanidins, effectively transferring some of the antioxidant nutrient value from the skin to the flesh of the root.

7.6   Genomic Tools for Nutritional and Cooking Quality Traits

Genomic tools are essential for identifying the genes and genetic variations that underlie nutritional quality traits in sweetpotato. These tools can help breeders to develop new varieties with improved nutritional and cooking quality and can also help researchers to better understand the biology of sweetpotato and the mechanisms that underlie these traits.

Reference Genome: Breeding for desired quality attributes in sweetpotato is challenging since traits that are of economic importance are often positively and negatively correlated. The possibility of pleiotropy in set of economic traits is inevitable. Identifying haplotypes which control traits of economic importance will help in facilitating selection decisions both at the phenotypic and molecular levels. Sequencing of the sweetpotato genome to identify genes associated with specific nutritional traits like beta-carotene (vitamin A precursor), starch and sugar content, iron uptake, and zinc accumulation is crucial. Genomic tools like a reference genome for cultivated sweetpotato is now available to facilitate genome enabled breeding for nutritional traits. Two diploid wild relatives of cultivated sweetpotato, *Ipomoea trifida* and Ipomoea triloba, have been sequenced and released now widely used as reference sequences in whole-genome studies (Wu et al. 2018). Comparative and phylogenetic analyses using these reference genomes provide insights into the ancient whole-genome triplication history of the genus Ipomoea. Researchers can now explore evolutionary relationships within the Batatas complex, which includes sweetpotato. By resequencing data from 16 genotypes widely used in African breeding programs, genes and alleles associated with carotenoid biosynthesis in sweetpotato storage roots have been identified. Genome browser for this haplotype-resolved chromosome-scale genome

Table 7.1 An example of sensory attributes constituting the lexicon for evaluation of cooked sweetpotato by a trained descriptive sensory panel

Assessment method	Descriptors	Simplified definition	Scale range
Aroma			
Once sample is received, slightly unwrap it, observe aroma with a single short whiff, close the foil, and mark your scores on the aroma scales	Sweetpotato	Smell of cooked sweetpotato	0 = none to 10 = very strong
	Caramel	Smell of burnt sugar or molasses (*sukaali gulu*)	0 = none to 10 = very strong
	Pumpkin	Smell of cooked pumpkin	0 = none to 10 = very strong
	Off-odor	"Unusual smells in sweetpotato including potato, boiled beans, amaranth, herbal, floral, and pungent/acidic/rotting sweetpotato"	0 = none to 10 = very strong
Appearance			
Visually assess the outer surface of the sweetpotato for orange color intensity	Orange color intensity	Intensity of orange color across the surface of the sample	0 = white, 1 = cream, 3 = yellow, 5 = yellow orange, 8 = orange, 10 = deep orange
Observe the cross-sectional cut for uniformity of color, degree of translucency and fibrousness	Uniformity of color	Evenness of color distribution across sample surface	0 = highly variable to 10 = consistent throughout
	Translucency	Quality of an object to allow light to pass through it but does not allow images to be distinguished such as a slice of steamed cucumber	0 = 100% chalky/opaque to 10 = 100% translucent
	Fibrous appearance	Presence of visible strings within sample mass	0 = none to 10 = extremely fibrous
Flavor			
Take a portion of the sample and chew slowly to score the intensity of the flavors	Sweetpotato	Intensity of the flavor of cooked sweetpotato	0 = none to 10 = very strong
	Pumpkin	Intensity of the flavor of cooked pumpkin	0 = none to 10 = very strong
	Cooked carrot	Intensity of the flavor of cooked carrot	0 = none to 10 = very strong
	Floral	Intensity of the flavor of flowers	0 = none to 10 = very strong
Take a portion of the sample and chew slowly to score the intensity of the basic tastes that you observe	Sweet	Taste of sugar	0 = not at all sweet to 10 = extremely sweet
	Bitter	Taste of quinine, strong coffee, *katunkuma* (*Solanum anguivi*), *nakati* (*Solanum aethiopicum*)	0 = not bitter 10 = extremely bitter
Texture in mouth			
Take a portion of sample and bite using front teeth (incisors) and assess fracturability	Fracturability	Ease with which sample breaks into distinct pieces when bitten between incisors	0 = easily deforms to 10 = easily fractures
Take another portion and bite using back teeth (molars) and assess hardness	Hardness in mouth	Amount of force required to compress product between molars	0 = extremely soft to 10 = hard

(continued)

Table 7.1 (continued)

Assessment method	Descriptors	Simplified definition	Scale range
While chewing (chew down), assess crunchiness and moisture in mass (3 chews)	Crunchiness	Production of low-pitched sound while chewing certain foods such as cooked carrot, cucumber	0 = not crunchy to 10 = extremely crunchy
	Moisture (in the mass)	Amount of moisture present in sample mass	0 = dry to 10 = extremely moist
After chewing 3 times, place sample between tongue and palate and assess crumbliness in mouth, adhesiveness, fibrousness, and smoothness	Crumbliness in mouth	Extent of powder like particles in sample mass	0 = not mealy to 10 = extremely mealy
	Adhesiveness	Amount of sample that adheres to oral surfaces	0 = none to 10 = very high
	Fibrousness	Presence of string like structures in mouth after chewing	0 = none to 10 = very high
	Smoothness	Degree of absence of grainy particles in mass	0 = grainy to 10 = very smooth
Take a portion of sample and chew until prompt to swallow to assess the rate of breakdown	Rate of breakdown	Number of chews required to masticate a sample until you can swallow it	0 = very slow to 10 = very fast
Texture by hand			
Press down the center of the sample to evaluate the force required to compress sample	Hardness by hand	Amount of force required to compress sample	0 = very soft, 5 = firm, 10 = very hard
Take a portion of sample and press between fore finger and thumb to assess moisture release	Moisture release	Attribute of food products to release moisture when pressure is applied such as cooked cucumber and French beans	0 = none to 10 = extremely moist
Attempt to make a ball from the sample to evaluate cohesiveness (moldability)	Cohesiveness (moldability)	Ease with which a ball like shape can be molded from sample	0 = falls apart to 10 = moldable
Rub a portion of sample between fingers to evaluate mealiness	Crumbliness (mealiness)	Ease with which sample breaks into small particles upon rubbing	0 = not mealy to 10 = extremely mealy

Source Nakitto et al. (2022)

assembly and annotation for different varieties of sweetpotato has been made available. It includes a set of search and query tools such as a BLAST server, genome browsers for two reference genomes, and gene report pages for all annotated genes in the species. This resource has bolstered efficient breeding of varieties with high provitamin A content.

Marker Development: Mapping complex traits genetically is by far the most expensive but also an important approach to identifying functional variants (Wallace et al. 2018). A number of quantitative trait loci have been identified for some of the nutritional quality traits using the Tanzania by Beauregard mapping population and its reciprocal cross in different environments in West Africa and United States of America. These two mapping populations are segregating for important quality attributes such as storage root dry matter, starch, β-carotene and sugar contents. Associating the phenotype and genotypes of clones in these two mapping populations in different environments enabled scientist to better understand the genetic architecture of quality attributes (Amankwaah 2019). Gemenet et al. 2020b using a biparental mapping population generated from a cross between an orange-fleshed and a non-orange-fleshed sweetpotato variety, identified two major QTLs located on linkage group (LG) three (LG3) and twelve (LG12) affecting starch, β-carotene, and their correlated traits, dry matter and flesh color. (Gemenet et al. 2020b). Some of the QTL discovered for nutritional quality have been hypothesized to be associated with important candidate genes in sweetpotato which could be targeted in improving cell wall structure, texture and flavor aside nutritional quality attributes. Based on the identified genes, scientists develop DNA markers. These markers function as flags, indicating the presence of genes linked to desired nutritional qualities. QTLs identified mentioned earlier were differentially expressed in Beauregard and Tanzania storage roots. It was reported that the two QTLs detected acted in a cis and trans manner to inhibit starch biosynthesis in amyloplasts and enhance chromoplast biogenesis, carotenoid biosynthesis,

and accumulation in OFSP. Breeders use these markers to screen large numbers of sweetpotato seedlings. Plants with the desired genetic markers are more likely to have improved nutritional content, allowing for faster selection and breeding.

Genomic Selection (GS): Genomic selection is a promising approach to enhance the nutritional quality of sweetpotato. It involves developing models to predict genotypes with desirable traits, such as higher protein content, essential amino acids, vitamins, and minerals, through genetic markers. This method can significantly accelerate the breeding process for developing nutrient-rich sweetpotato varieties. Genomic selection is a promising approach to enhance the nutritional quality of sweetpotato. It involves developing identifying and selecting desirable traits, such as higher protein content, essential amino acids, vitamins, and minerals, through genetic markers. This method can significantly accelerate the breeding process for developing nutrient-rich crop varieties. Predicting the genetic value of the sweetpotato germinated seedlings with the help of genome-wide marker data to identify individuals of high nutritional quality prior to phenotypic assessment. Functional genomics: Understanding the functions of genes involved in nutrient biosynthesis and metabolism, which facilitate the manipulation of these pathways to improve nutrient content.

7.7 Challenges and Future Directions

Efforts to increase nutrition must be balanced with other traits, emphasizing the need for nutritional genomics to influence overall crop genotyping strategies. This raises difficulties because nutritional traits often associated with consumer benefits have quantitative inheritance and multifactorial genetic causation. The enormous genetic diversity in sweetpotato also presents challenges. Gaining access to genotypes representing the full range of nutritional phenotypes can be difficult, but the main challenge is to understand and develop crops with increased nutrition that are suited to the full range of environments where sweetpotato is grown in the developing world. This will need extensive research on genotype by environment by management interaction, and increased nutrition must be accompanied by maintenance or enhancement of productivity (Low et al. 2020). One of the enabling technologies for genetic enhancement of complex traits, including nutrition, is transgenesis, which often acts as a genetic "proof" of the identity of genes and their function. But transgenesis in many food crops, including sweetpotato, does not lead to commercialization and its adoption for crop improvement has been variable, despite offering significant benefits over alternative technologies such as marker-assisted selection. This is because regulatory, biosafety, and consumer acceptance barriers are high for transgenesis. Alternative strategies such as MAS must be used as a bridge to commercial biotech, and there is concern that misinterpretation of transgenesis will lead to rejection of all biotechnology, negating the potential benefits of genomics on nutritional enhancement. Efforts in genomics-assisted breeding are anticipated to promote increased nutritional quality of sweetpotato. Success would heighten the profile of sweetpotato as a health-promoting food and support the crop's contribution to food security in the developing world. However, success is not assured and there are a number of technical and commercialization challenges to be overcome.

References

Amankwaah VA (2019) Phenotyping and genetic studies of storage root chemistry traits in sweetpotato. North Carolina State University, Raleigh

Amankwaah VA, Williamson S, Reynolds R et al (2023) Development of NIRS calibration curves for sugars in baked sweetpotato. J Sci Food Agric 104:4801–4807. https://doi.org/10.1002/jsfa.12800

Bouis HE, Saltzman A (2017) Improving nutrition through biofortification: a review of evidence from HarvestPlus, 2003 through 2016. Glob Food Sec 12:49–58. https://doi.org/10.1016/j.gfs.2017.01.009

Bouis HE, Welch RM (2010) Biofortification: a sustainable agricultural strategy for reducing micronutrient malnutrition in the global south. Crop Sci 50:20–32

Chen BH, Chen YY (1993) Stability of chlorophylls and carotenoids in sweetpotato leaves during microwave cooking. J Agric Food Chem 41:1315–1320

Da Silva Pereira GS, Amankwaah VA, Ketavi M et al (2023) Sweetpotato: nutritional constituents and genetic composition. Compendium of crop genome designing for nutraceuticals. Springer, Singapore, pp 1–43

de Albuquerque TMR, Sampaio KB, de Souza EL (2019) Sweetpotato roots: unrevealing an old food as a source of health promoting bioactive compounds—a review. Trends Food Sci Technol 85:277–286. https://doi.org/10.1016/j.tifs.2018.11.006

FAOSTAT (2022) World food and agriculture: statistical yearbook 2022. FAO, New York

Gemenet DC, da Silva PG, De Boeck B et al (2020a) Quantitative trait loci and differential gene expression analyses reveal the genetic basis for negatively associated β-carotene and starch content in hexaploid sweetpotato [*I. batatas* (L.) Lam.]. Theor Appl Genet 133:23–36. https://doi.org/10.1007/s00122-019-03437-7

Gemenet DC, Kitavi MN, David M, Ndege D, Ssali RT, Swanckaert J, Makunde G, Yencho GC, Gruneberg W, Carey E, Mwanga RO, Andrade MI, Heck S, Campos H (2020b) Development of diagnostic SNP markers for quality assurance and control in sweetpotato [*Ipomoea batatas* (L.) Lam.] breeding programs. PLOS ONE 15(4):e0232173. https://doi.org/10.1371/journal.pone.0232173

Grüneberg WJ, Ma D, Mwanga ROM et al (2015) Advances in sweetpotato breeding from 1992 to 2012. Potato and sweetpotato in Africa: transforming the value chains for food and nutrition security. CABI, New York, pp 3–68

Gruneberg WJ, Eyzaguirre R, Diaz F, De Boeck B, Espinoza J, Mwanga ROM, Swanckaert J, Dapaah H, Andrade M, Makunde G, Tumwegamire S, Agili S, Ndingo-Chipungu FP, Attaluri S, Kapinga R, Nguyen T, Kaiyung X, Tjintokohadi K, Ssali RT, Carey T, Low J (2019) Procedures for the evaluation of sweetpotato trials. Manual. Lima (Peru). International Potato Center (CIP), p 86

Haas JD, Beard JL, Murray-Kolb LE, del Mundo AM, Felix A, Gregorio GB (2005) Iron-biofortified rice improves the iron stores of nonanemic filipino women12. J Nutr 135(12):2823–2830. https://doi.org/10.1093/jn/135.12.2823

IFT (Institute of Food Technologists) (2007) Sensory evaluation methods. The Society for the Food Technologists, Chicago, IL

Jackson MJ (1997) The assessment of bioavailability of micronutrients: introduction. Eur J Clin Nutr 51:S1–S2

Jenkins M, Shanks CB, Brouwer R, Houghtaling B (2018) Factors affecting farmers' willingness and ability to adopt and retain vitamin A-rich varieties of orange-fleshed sweetpotato in Mozambique. Food Sec 10:1501–1519. https://doi.org/10.1007/s12571-018-0845-9

Joanna L, Inés MA, Esteban V, Gustavo R, Patricia A, Mariana R et al (2019) Integration of sensory analysis into plant breeding: a review. Agrociencia Uruguay 23:1–15

Kays SJ, Wang Y (2002) Sweetpotato quality: its importance, assessment and selection in breeding programs. Acta Hortic 187–193. https://doi.org/10.17660/ActaHortic.2002.583.21

Kemp SE, IFST PFSG Committee (2008) Application of sensory evaluation in food research. Int J Food Sci Technol 43:1507–1511

Kemp SE, Ng M, Hollowood T, Hort J (2018) Introduction to descriptiveanalysis. In: Kemp SE, Hort J, Hollowood T (eds) Descriptive analysis in sensory evaluation. Wiley, Hoboken, pp 3–39

Khan Z, Takemura M, Maoka M, Hattan T, Otani M, Misawa N (2022) Molecular breeding of sweetpotato carotenoids. IntechOpen. https://doi.org/10.5772/intechopen.101849

Kitahara K, Nakamura Y, Otani M et al (2017) Carbohydrate components in sweetpotato storage roots: their diversities and genetic improvement. Breed Sci 67:62–72. https://doi.org/10.1270/jsbbs.16135

Low JW, Ortiz R, Vandamme E, Andrade M, Biazin B, Grüneberg WJ (2020) Nutrient-dense orange-fleshed sweetpotato: advances in drought-tolerance breeding and understanding of management practices for sustainable next-generation cropping systems in Sub-Saharan Africa. Front Sustain Food Syst 4. https://doi.org/10.3389/fsufs.2020.00050

Mudege NN, Kebaara K, Mukewa E (2019) Effects of commercialization of sweetpotato on gender relations and wellbeing among smallholder farmers: technical workshop to review study findings and develop recommendations for improved programming

Mwanga ROM, Swanckaert J, da Silva PG et al (2021a) Breeding progress for vitamin A, iron and zinc biofortification, drought tolerance, and sweetpotato virus disease resistance in sweetpotato. Front Sustain Food Syst 5:6674. https://doi.org/10.3389/fsufs.2021.616674

Mwanga ROM, Mayanja S, Swanckaert J, Nakitto M, Felde T, Gruneberg W et al (2021b) Development of a food product profile for boiled and steamed sweetpotato in Uganda for effective breeding. Int J Food Sci 56:1385–1398

Nakamura Y, Ohara-Takada A, Kuranouchi T, Masuda R, Katayama K (2014) Mechanism for maltose generation by heating in the storage roots of sweetpotato cultivar "Quick Sweet" containing starch with a low pasting temperature. Nippon Shokuhin Kagaku Kogaku Kaishi 61:577–585

Nakatumba-Nabende J, Babirye C, Tusubira JF, Mutegeki H, Nabiryo AL, Murindanyi S, Katumba A, Nantongo J, Sserunkuma E, Nakitto M, Ssali R, Makunde G, Moyo M, Campos H (2023) Using machine learning for image-based analysis of sweetpotato root sensory attributes. Smart Agricult Technol 16:291. https://doi.org/10.1016/j.atech.2023.100291

Nakitto M, Johanningsmeier SD, Moyo M, Bugaud C, de Kock H, Dahdouh L, Forestier-Chiron N, Ricci J, Khakasa E, Ssali RT, Mestres C (2022) Sensory guided selection criteria for breeding consumer-preferred sweetpotatoes in Uganda. Food Qual Prefer 101:104628. https://doi.org/10.1016/j.foodqual.2022.104628

Neela S, Fanta SW (2019) Review on nutritional composition of orange-fleshed sweetpotato and its role in management of vitamin A deficiency. Food Sci Nutr 7:1920–1945. https://doi.org/10.1002/fsn3.1063

Ravi V, Chakrabarti SK, Makeshkumar T, Saravanan R (2014) Molecular regulation of storage root formation and development in sweetpotato. Hortic Rev Am Soc Hortic Sci 42:157–208. https://doi.org/10.1002/9781118916827.ch03

Ru L, Chen B, Li Y, Wills RBH, Lv Z, Lu G, Yang H (2021) Role of sucrose phosphate synthase and vacuolar invertase in postharvest sweetening of immature sweetpotato tuberous roots (*I. batatas* (L.) Lam Cv 'Xinxiang'). Sci Hortic 282:110007

Sanoussi AF, Adjatin A, Dansi A, Adebowale A, Sanni LO, Sanni A (2016) Mineral composition of ten elites sweetpotato (*I. batatas* [L.] Lam.) Landraces of Benin. Int J Curr Microbiol Appl Sci 5(1):103–115

Shahzad R, Jamil S, Ahmad S et al (2021) Biofortification of cereals and pulses using new breeding techniques: current and future perspectives. Front Nutr 8:721728. https://doi.org/10.3389/fnut.2021.721728

Shimamura T, Sumikura Y, Yamazaki T, Tada A, Kashiwagi T, Ishikawa H, Ukeda H (2014) Applicability of the DPPH Assay for evaluating the antioxidant capacity of food additives: inter-laboratory evaluation study. Anal Sci 30:717–721

Takiko S, Motoyasu O, Tatsuro H, Sun-Hyung K (2006) Increase of amylose content of sweetpotato starch by RNA interference of the starch branching enzyme II gene (IbSBEII). Plant Biotechnol 23(1):85–90. https://doi.org/10.5511/plantbiotechnology.23.85

Tomlins K, Ndunguru G, Stambul K et al (2007) Sensory evaluation and consumer acceptability of pale-fleshed and orange-fleshed sweetpotato by school children and mothers with preschool children. J Sci Food Agric 87:2436–2446. https://doi.org/10.1002/jsfa.2931

Truong VD, Avula RY, Pecota KV, Yencho GC (2018) Sweetpotato production, processing, and nutritional quality. Handbook of vegetables and vegetable processing. Wiley, Amsterdam, pp 811–838

Tumwegamire S, Rubaihayo PR, LaBonte DR et al (2011a) Genetic diversity in white- and orange-fleshed sweetpotato farmer varieties from east Africa evaluated by simple sequence repeat markers. Crop Sci 51:1132–1142. https://doi.org/10.2135/cropsci2010.07.0407

Tumwegamire S, Kapinga R, Rubaihayo PR, LaBonte DR, Grüneberg WJ, Burgos G, Felde TZ, Carpio R, Pawelzik E, Mwanga RO (2011b) Evaluation of dry matter, protein, starch, sucrose, β-carotene, iron, zinc, calcium, and magnesium in east African sweetpotato [*I. batatas* (L.)

Lam] germplasm. HortScience Horts 46(3):348–357. https://doi.org/10.21273/HORTSCI.46.3.348

Wallace JG, Rodgers-Melnick E, Buckler ES (2018) On the road to breeding 4.0: unravelling the good, the bad and the boring of crop quantitative genomics. Annu Rev Genet 52:421–444

Welch RM, Graham RD (2004) Breeding for micronutrients in staple food crops from a human nutrition perspective. J Exp Bot 55:353–364

Woolfe J (1992) Sweetpotato: an untapped food resource. Cambridge University Press and the International Potato Center (CIP), Cambridge

Wu S, Lau KH, Cao Q, Hamilton JP, Sun H, Zhou C, Eserman L, Gemenet DC, Olukolu BA, Wang H, Crisovan E, Godden GT, Jiao C, Wang X, Kitavi M, Manrique-Carpintero N, Vaillancourt B, Wiegert-Rininger K, Yang X, Bao K, Schaff J, Kreuze J, Gruneberg W, Khan A, Fei Z (2018) Genome sequences of two diploid wild relatives of cultivated sweetpotato reveal targets for genetic improvement. Abstr Nat Commun 9(1). https://doi.org/10.1038/s41467-018-06983-8

Molecular Breeding of Carotenoids in Sweetpotato

8

Mercy Kitavi and C. Robin Buell

Abstract

This chapter overviews molecular breeding efforts focused on enhancing carotenoid content in sweetpotato. Sweetpotato is a widely cultivated crop known for its adaptability to diverse climates and soil conditions, making it a staple food in many regions worldwide. Sweetpotato also offers notable nutritional and health benefits, owing to its rich content of essential vitamins, minerals, and antioxidants. Of particular interest is β-carotene, a precursor of vitamin A, abundant in orange-fleshed sweetpotato varieties. A vital nutrient for human health, β-carotene serves as a key focus in efforts to enhance the nutritional quality of sweetpotato. Identification and expression of carotenoid biosynthesis genes provide valuable insights into the genetic mechanisms underlying carotenoid accumulation and starch metabolism in sweetpotato storage roots. Through breeding, researchers can develop sweetpotato varieties with elevated β-carotene content, improving their nutritional value and health-promoting properties. Future directions in molecular breeding of carotenoids in sweetpotato will involve the integration of advanced genetic tools and technologies to accelerate trait improvement and meet the evolving nutritional needs of diverse populations. This, in combination with other tools such as gene editing, holds promise for enhancing β-carotene content in sweetpotato to address malnutrition and promote public health initiatives globally.

Keywords

Sweetpotato · Molecular breeding · Nutrition · Carotenoids

8.1 Sweetpotato and Its Production

Ipomoea, the largest genus in the Convolvulaceae family, encompasses 600–700 species, including *I. batatas* (L.) Lam. (sweetpotato), which is extensively cultivated worldwide as a food crop (Hirakawa et al. 2015; Austin et al. 2015). Sweetpotato ranks seventh globally among the most valuable food crops, following wheat, rice, maize, potato, barley, and cassava

M. Kitavi (✉)
Research Technology Support Facility (RTSF), Genomics Core, Michigan State University, East Lansing, MI 48824, USA
e-mail: kitavi.mercy@gmail.com; kitavime@msu.edu

C. R. Buell
Center for Applied Genetic Technologies, Institute of Plant Breeding, Genetics and Genomics, and Department of Crop and Soil Sciences, University of Georgia, Athens, GA 30602, USA

G. C. Yencho et al. (eds.), *The Sweetpotato Genome*, Compendium of Plant Genomes, https://doi.org/10.1007/978-3-031-65003-1_8

(CIP 2020; https://cipotato.org/sweetpotato/sweetpotato). Cultivated in over 115 countries, sweetpotato boasted an annual production of 91.8 million metric tons in 2022 (FAOSTAT 2022). Asia leads in sweetpotato production, accounting for 82 million tons (81.4%), primarily driven by China, followed by Africa with 17 million tons (15.2%) (FAOSTAT 2022). Notably in Africa, Malawi, Nigeria, United Republic of Tanzania, and Uganda rank among the top most sweetpotato producers after China (FAOSTAT 2022), highlighting its significance as a secondary staple food root crop, alongside cassava, and its substantial role in human diets (van Jaarsveld et al. 2005; Low et al. 2009).

Because of its ability to thrive in nutrient-poor soils with minimal input, sweetpotato is cultivated across diverse agroecological and microclimatic zones, spanning from tropical to temperate climates (Niringiye et al. 2014). Sweetpotato plays multiple roles in the global food system, each of which carries significant implications for meeting food needs, alleviating poverty, and enhancing food security (Low et al. 2017; El-Sheikha and Ray 2017). The roots of sweetpotato possess higher levels of carbohydrates, minerals, and protein compared to other tropical root and tuber crops (Ji et al. 2015).

While the protein content of sweetpotato, like most tropical root and tuber crops, is relatively low (around 2%), it surpasses that of cassava and plantain (Woolfe 1992).

8.2 Nutritional and Health Benefits of Sweetpotato

Sweetpotato roots are rich in secondary metabolites that offer significant nutritional benefits and exhibit remarkable sensory versatility, encompassing taste, texture, and flesh color. The flesh color of the roots range from white to cream, yellow, orange, and purple (Fig. 8.1). In sub-Saharan Africa, selective breeding efforts have resulted in the development of varieties characterized by storage roots with high dry matter (>25%) and varying flesh colors, including white, cream, and yellow. These varieties also boast higher starch content and a "mealy firm" texture upon cooking. Starch serves as the primary carbohydrate in sweetpotato storage roots, with the composition, size, and shape of starch granules playing pivotal roles in determining eating quality (Kitahara et al. 2017; Reeve 1967; Lv et al. 2019).

Sweetpotato cultivars featuring orange flesh color (referred to as orange fleshed; OFSP) are

Fig. 8.1 Sweetpotato flesh color range from white to yellow, orange, and purple. Photo Credit: Mercy Kitavi, 2019

renowned for their rich content of non-digestible dietary fiber, minerals, vitamins, and antioxidants (Neela and Fanta 2019; Dako et al. 2016). Notably, they exhibit high levels of β-carotene but relatively low dry matter content (18–25%). These varieties are typically characterized by a sweet flavor and moist texture post-cooking, making them a popular choice commercially, particularly in the USA (Islam et al. 2016; Grace et al. 2014; Liao et al. 2008).

Purple-fleshed sweetpotatoes, known for their abundance of anthocyanins, are a specialty variety particularly popular in Asia. These sweetpotatoes display an appealing purple-red hue and are characterized by high levels of anthocyanins, total phenols, and antioxidant activity (Steed and Truong 2008; Yoshinaga et al. 1999). Across different sweetpotato varieties, at least 27 anthocyanin pigments have been identified (He et al. 2016; Wang et al. 2018; Lee et al. 2013). Compared to orange-fleshed sweetpotatoes, purple-fleshed varieties exhibit significantly higher anthocyanin content (Kurnianingsih et al. 2020), akin to other anthocyanin-rich crops such as blueberries, blackberries, cranberries, and grapes (Bridgers et al. 2010). Furthermore, purple sweetpotatoes serve as an economical source of natural anthocyanin pigments (Jansen and Flamme 2006). The anthocyanins present in purple-fleshed sweetpotatoes encompass various chemical structures, primarily cyanidins and peonidins; these acylated forms offer heat and light stability properties alongside antioxidant activity (Mu et al. 2021), making them desirable natural pigments for food additives (Odake et al. 1994; Xu et al. 2015). Additionally, purple-fleshed sweetpotato varieties exhibit higher dry matter content (up to 38.96%) compared to OFSP, with a negative correlation observed between anthocyanin levels and water content (Steed and Truong 2008).

8.3 β-Carotene is an Important Vitamin a Source for Humans

Carotenoids play diverse roles in human health, ranging from acting as antioxidants to supporting vision and immune function (Eggersdorfer and Wyss 2018). Dietary compounds with vitamin A activity encompass both preformed all-trans-retinol (referred to here as retinol for simplicity) and retinyl esters, along with provitamin A carotenoids like β-carotene or β-cryptoxanthin (Fraser and Bramley 2004; Krinsky and Johnson 2005; von Lintig 2012; Scott and Ewell 1992). Apart from certain aphids that naturally produce the carotenoid torulene (Moran and Jarvik 2010), animals lack the ability to synthesize these essential nutritional molecules de novo and therefore depend on dietary sources of this vital vitamin (Goodwin 1984).

The primary precursors of vitamin A (VA) in the human body include β-carotene, α-carotene, and β-cryptoxanthin (Arscott and Tanumihardjo 2010). The β-carotene (BC) found in OFSP plays a significant role as a long-term food-based strategy for combating vitamin A deficiency, as evidenced by recent studies in Africa (Low et al. 2017; World Food Prize Foundation 2016). Dietary carotenoids such as α-carotene and β-carotene have beneficial effects on human health, including antioxidant activity, supporting immune function, and reducing the risk of chronic diseases (Eggersdorfer and Wyss 2018; Fiedor and Burda 2014). Furthermore, lutein and zeaxanthin, both carotenoids, serve as macular pigments that aid in protecting the eyes and reducing the risk of age-related macular degeneration and cataracts (Sauer et al. 2019). Many OFSP varieties contain up to 276.98 μg of β-carotene per gram of fresh weight (Low et al. 2007; Kang et al. 2017; Grune et al. 2010). Utilizing OFSP can help improve vitamin A status and enhance the bioavailability of various micronutrients such as iron, zinc, calcium, and magnesium, thereby reducing the risk of vitamin A deficiency (Islam et al. 2016; Vimala et al. 2011; Gurmu et al. 2014). Moreover, β-carotene, as the provitamin A carotenoid with antioxidant properties and the highest vitamin A activity, has been associated with boosting the immune system and reducing the risk of cancer (Fiedor and Burda 2014).

In many developing nations, sweetpotato serves as a secondary staple food, bridging nutritional gaps and bolstering the intake of essential vitamins and minerals, particularly in combating vitamin A deficiency among children, pregnant women, and lactating mothers (Han et al. 2022; Low and Thiele 2020). Orange-fleshed sweetpotato varieties, rich in β-carotene, have proven successful in providing provitamin A biofortification in sub-Saharan Africa (SSA) (Low et al. 2009, 2017; Neela and Fanta 2019). However, sweetpotato cultivation in SSA has traditionally centered on varieties preferred for their high starch content, such as white and yellow-fleshed types, which, despite their high dry matter content, offer lower nutritional value (Low et al. 2017). These varieties have been specifically chosen for their elevated starch levels. Moreover, local consumers favor starchy sweetpotato varieties for their distinct textural attributes after cooking (Jenkins et al. 2018), influenced by factors like texture and sweetness, which are contingent on the composition and quantity of carbohydrates, including cellulose, hemicellulose, pectin, starch, and sugars (Reeve 1967).

8.4 Carotenoid Biosynthetic Pathway in Higher Plants

Carotenoids are tetraterpene pigments that derive their name from the carrot (*Daucus carota*), a plant renowned for accumulating high levels of these pigments in its roots. While carotenoids commonly impart color to flowers, fruits, and seeds in plants (Hirschberg 2001), their accumulation in underground organs like tubers and roots represents an exception. Primarily, carotenoids serve in light-harvesting processes by safeguarding the plant's photosynthetic machinery against photo-oxidative damage (Zakar et al. 2016). Nature consists of hundreds of carotenoid structures, broadly classified into carotenes (hydrocarbons capable of cyclization at one or both ends of the molecule) and xanthophylls (oxygenated derivatives of carotenes) (Ruiz-Sola and Rodríguez-Concepción 2012). Carotenes, predominantly β-carotene, abound in the photosystem reaction centers,

while xanthophylls are most prevalent in the light-harvesting complexes, (Davison et al. 2002; Pogson et al. 1998).

The carotenoid biosynthetic pathway (Fig. 8.2a) has been extensively elucidated in numerous plant species, including Arabidopsis (*Arabidopsis thaliana*) (Ruiz-Sola and Rodríguez-Concepción 2012), tomato (*Solanum lycopersicum*) (Bramley 2002), maize (*Zea mays*) (Vallabhaneni and Wurtzel 2009), and rice (*Oryza sativa*) (Beyer et al. 2002). Carotenoids in higher plants are synthesized through the condensation of geranylgeranyl pyrophosphate (GGPP) from the methylerythritol 4-phosphate (MEP) pathway into phytoene (Auldridge et al. 2006). This initial committed step, catalyzed by the enzyme phytoene synthase (PSY), is considered the principal bottleneck in the carotenoid pathway (Cazzonelli and Pogson 2010; Sandmann et al. 2006). Subsequently, through desaturation and isomerization processes involving enzymes like phytoene desaturase (PDS), 15-cis-ζ-carotene isomerase (Z-ISO), ζ-carotene desaturase (ZDS), and carotenoid isomerase (CRTISO), the plant carotenoid backbone is synthesized (Britton 1995), ultimately leading to the formation of the linear carotenoid lycopene, which imparts a red color (Burton and Ingold 1984; Ruiz-Sola and Rodríguez-Concepción 2012). The first divergence in the pathway occurs when lycopene undergoes cyclization, catalyzed by lycopene β-cyclase (LCY-β) and/or lycopene ε-cyclase (LCY-ε), resulting in the production of orange α-carotene and β-carotene, representing the α- and β-branches of the pathway, respectively. α-carotenoids possess one ß ring and one ε ring (α-carotene), whereas ß-carotenoids feature two ß rings (Chen et al. 2010). α-Carotene, β-carotene, and β-cryptoxanthin are considered provitamin A carotenoids, as they can be converted by the body into retinol. Further modifications of carotenes and xanthophylls lead to the synthesis of various species-specific carotenoids (Giuliano 2014, 2017). Notably, lutein, zeaxanthin, and lycopene are non-provitamin A carotenoids and cannot be converted to retinol.

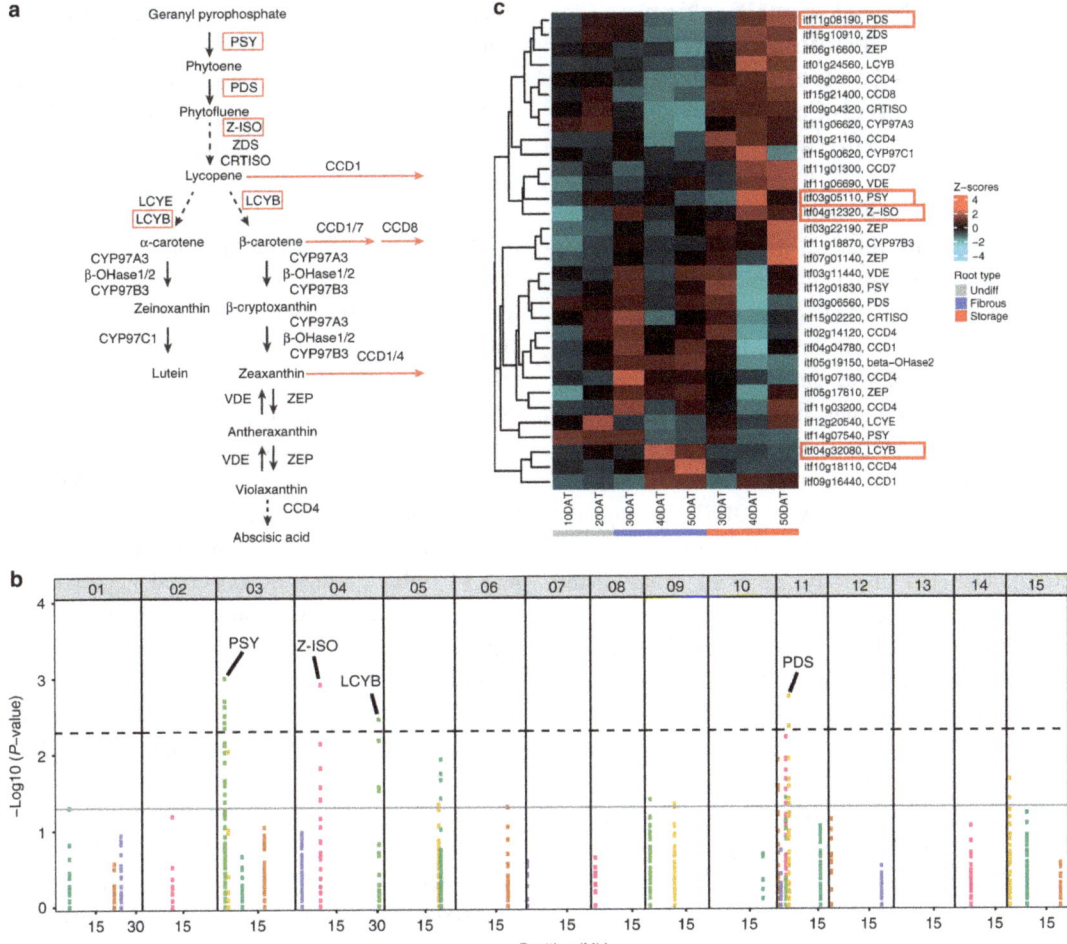

Fig. 8.2 **a** Carotenoid biosynthesis and degradation pathways. Multi-step reactions are shown by dashed arrows and red arrows indicate degradation reactions. Genes in red rectangles were those found with significantly enriched SNPs at $p < 0.005$ (Fisher's exact test). PSY, phytoene synthase; PDS, phytoene desaturase; Z-ISO, ζ-carotene isomerase; ZDS, ζ-carotene desaturase; CRTISO, carotenoid isomerase; LCYE, lycopene ε-cyclase; LCYB, lycopene β-cyclase; β-OHase, β-ring hydroxylase; CYP, cytochrome P450; VDE, violaxanthin de-epoxidase; ZEP, zeaxanthin epoxidase; CCD, carotenoid cleavage dioxygenase. **b** Allele frequencies ($p = 0.005$; dotted lines and $p = 0.05$; solid lines) for putative carotenoid biosynthesis loci between orange and white-fleshed accessions. SNPs from the same gene in each chromosome are indicated by a different color. **c** Expression profiles of genes involved in carotenoid biosynthesis in different types of roots of orange-fleshed cultivar, "Beauregard" during development. Reprinted from "Genome sequences of two diploid wild relatives of cultivated sweetpotato reveal targets for genetic improvement" Wu et al. (2018). Copyright 2018 by Creative Commons CC BY Reprinted with permission

8.5 Functional Identification of Carotenoid Biosynthesis Genes Controlling Carotenoid and Starch Content in Sweetpotato

Access to genome sequences of a wild diploid sweetpotato relative, *Ipomoea trifida* along with other *Ipomoea* genomes (https://ipomoea-genome.org/) (Wu et al. 2018; Wadl et al. 2018; da Silva Pereira et al. 2020), has facilitated the understanding of the genetic architecture and identification of genes involved in starch and carotenoid biosynthesis (Gemenet et al. 2020). Varietal flesh color is correlated with the amount of β-carotene content with OFSP having the highest amount of β-carotene (Gemenet et al. 2020). Using a diversity panel of orange- and white-fleshed sweetpotato, possible loci (single nucleotide polymorphisms) involved in the accumulation of β-carotene in OFSP were identified in key carotenoid biosynthetic genes [phytoene synthase (PSY; itf03g05110), phytoene desaturase (PDS; itf11g08190) and ζ-carotene isomerase (Z-ISO; itf04g12320) (Fig. 8.2b)]. These loci can be used as targets for marker-assisted selection of crosses for beta carotene. A comparison on root types (Fig. 8.2c) showed an upregulation of the PSY only in the orange-fleshed storage roots suggesting its involvement in conferring the orange color in OFSP storage roots (Gemenet et al. 2020). An expression bias in the PSY "orange" alleles was observed in OFSP during the later stages of storage root development, suggesting a correlation with carotenoid accumulation (Gemenet et al. 2020). Likewise, in the initial month of growth, young carrot roots appear pale but gradually accumulate carotenoids, reaching peak levels around three months later, just prior to the completion of secondary growth (Baranska et al. 2006).

New linkage and quantitative trait loci (QTL) mapping methods for polyploids (Mollinari et al. 2020; da Silva Pereira et al. 2020) have aided in identification of a QTL (Gemenet et al. 2020) unraveling the genetic basis for the negative association between β-carotene and starch.

Carotenoid accumulation was observed in both storage and fibrous roots of OFSP, whereas it was not reported in white-fleshed varieties (Gemenet et al. 2020). A major QTL co-localized on LG3 and LG12 of the integrated genetic map explained variation in dry matter (DM), starch content, β-carotene levels, and flesh color (FC) (Fig. 8.3). Both parental lines contributed major alleles with comparable effects on traits located on LG12, while only the OFSP parent, Beauregard, exhibited significant allelic effects on traits at the LG3 QTL. Additionally, dry matter (DM), starch, β-carotene (BC), and flesh color (FC) traits are influenced by additive allele effects. The same contributing haplotypes responsible for reducing DM and starch content were also associated with an increase in BC and FC, elucidating the observed negative correlation between starch and BC in sweetpotato, a phenomenon akin to that observed in cassava (Rabbi et al. 2017).

The contrasting traits exhibited by the two parents suggest that the interaction of alleles between the two QTLs dictates the presence or absence of β-carotene accumulation in sweetpotato storage roots. Additionally, the rate-limiting PSY gene (itf03g05110) in carotenoid biosynthesis was found to be positioned between the two co-localized QTL peaks on LG3, specifically at 2,994,719 bp (associated with β-carotene and flesh color) and 3,185,578 bp (linked with dry matter and starch) in *I. trifida*. The starch gene sucrose synthase (SuSY; itf03g05100) exhibited early expression during root initiation (10–20 days after transplanting) in both white and orange-fleshed cultivars. However, it continued to be expressed in white-fleshed sweetpotato beyond 50 days after transplanting. Moreover, the *PSY* gene and *SuSY* genes are situated within a 12.2 kb region with no intervening genes.

The orange (OR) protein plays a pivotal role in regulating carotenoid accumulation through post-transcriptional regulation of PSY, facilitating the formation of carotenoid-sequestering structures, and preventing carotenoid degradation (Chayut et al. 2017; Zhou et al. 2015).

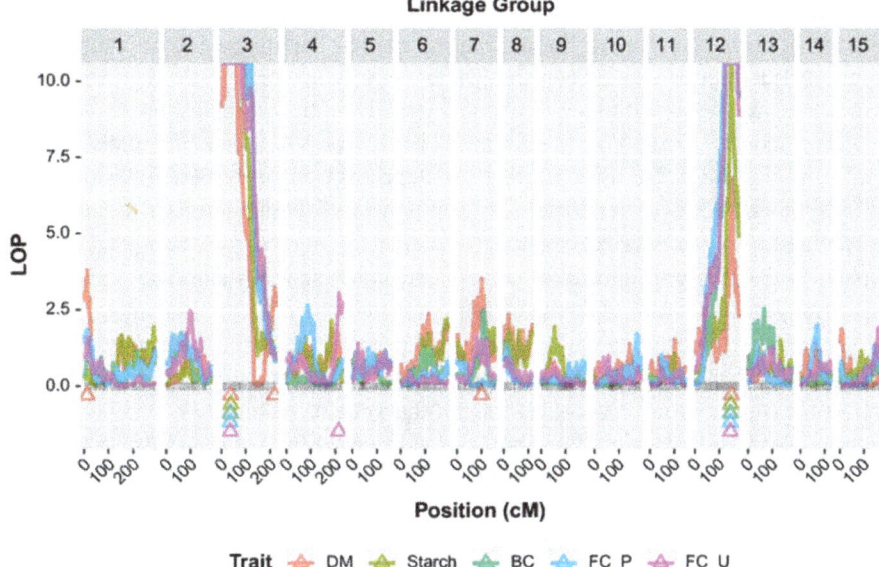

Fig. 8.3 Quantitative trait loci (QTL) profiles as LOP = −log10(*p*) for β-carotene (BC), dry matter (DM), starch, and flesh color (FC in Peru and Uganda; denoted by P and U, respectively) along the genetic map of a biparental (Beauregard × Tanzania) hexaploid sweetpotato population. QTL location peaks are marked by triangles. Reprinted from "Quantitative trait loci and differential gene expression analyses reveal the genetic basis for negatively associated β-carotene and starch content in hexaploid sweetpotato [*Ipomoea batatas* (L.) Lam.]" Gemenet et al. (2020). Copyright 2020 by Creative Commons CC BY. Reprinted with permission

Intriguingly, the Or gene (itf12g24270), situated 5.7 kb from the LG12 QTL peak, has been linked with β-carotene accumulation in sweetpotato (Gemenet et al. 2020). Evidence suggests that Or modulates PSY, enabling the transformation of amyloplasts into chromoplasts in various crops such as cauliflower (Lu et al. 2006), tomato (Yazdani et al. 2019), Arabidopsis (Bai et al. 2016), corn (Berman et al. 2017), melon (Tzuri et al. 2015), and sweetpotato (Kim et al. 2013; Park et al. 2015), including purple-fleshed sweetpotato cultivars. Post-translational mechanisms governing OR and PSY protein stability contribute to increased carotenoid levels in yellow cassava (Jaramillo et al. 2022).

8.6 Breeding for β-Carotene Content and Future Directions

Although research has prioritized the fortification of sweetpotato with provitamin A carotenoids (Low and Thiele 2020; Low et al. 2017), the mechanisms underlying carotenoid accumulation in underground storage roots remain poorly understood across various crops, including sweetpotato (Carvalho et al. 2016). Carotenoids are synthesized within various plastids, including proplastids, amyloplasts, etioplasts, chloroplasts, and chromoplasts (Jarvis and López-Juez 2013). Among these, all plastids except proplastids play crucial roles in regulating carotenogenic activity, carotenoid stability, and pigment diversity (Li et al. 2016). Amyloplasts are predominantly found in starchy organs such as wheat, rice, barley, and maize seeds, as well as potato tubers and cassava roots (Jarvis and López-Juez 2013). They primarily synthesize and accumulate carotenoids, particularly xanthophylls like lutein, zeaxanthin, and violaxanthin (Sun et al. 2018; Wurtzel et al. 2012; Wurtzel 2019). Several factors, including biosynthetic capacity, plastid ultrastructure, and metabolic channeling, may limit carotenoid biosynthesis and accumulation in amyloplasts. In contrast, chromoplasts exhibit superior capabilities for carotenoid sequestration and storage by forming carotenoid-lipoprotein sequestering substructures (Li and Yuan 2013). These substructures are proposed to act as a sink for sequestering excess carotenoids, ensuring stable storage and preventing an overload of the carotenoid biosynthetic pathway products.

Among root and tuber crops, sweetpotato distinguishes itself by its ability to induce the transformation of amyloplasts into crystalline-type carotenoid sequestration substructures known as amylochromoplasts (Drapal et al. 2022). This process alters both the capacity for carotenoid storage and biosynthesis, leading to increased accumulation of β-carotene (Zhang et al. 2014; Drapal et al. 2022). The observation of a mutually exclusive relationship between carotenoid accumulation and starch granule development in tobacco floral nectaries and carrot roots implies that increased carotenogenesis may act as a developmental cue, guiding the transition from amyloplasts to chromoplasts (Kim et al. 2010; Horner et al. 2007).

Moreover, alterations in chromoplast morphology observed in plants engineered for enhanced carotenoid production suggest an adaptation of cellular structures to facilitate the sequestration of newly synthesized carotenoids (Horner et al. 2007; Kim et al. 2010). However, the storage of carotenoids in modified amyloplasts leads to competition for carbon resources between starch and carotenoid biosynthesis, resulting in a negative correlation, as observed in sweetpotato (Gemenet et al. 2020; Yada et al. 2017). Similar phenomena have been documented in other crops, including citrus (Cao et al. 2015), potato (Mortimer et al. 2016; Fernandez-Orozco et al. 2013), and cassava (Olayide et al. 2020).

Biofortification of sweetpotato landraces has been a continuous process leading to improved carotenoid content attributable to the adoption of advanced breeding techniques including the screening of large numbers of genotypes for nutritional quality, agronomic traits, yield traits, and the selection of progenies with the optimal traits for further breeding (Yada et al. 2017; Gemenet et al. 2020). Efforts to biofortify sweetpotato have focused on increasing β-carotene content and improving

organoleptic qualities of commonly consumed varieties. Replacement of white-fleshed sweetpotato with orange-fleshed varieties has benefited ~50 million children <6 years of age at risk of VA deficiency (Low et al. 2017; van Jaarsveld et al. 2005). Furthermore, OFSP clones are being selected for other health traits such as increased Fe and Zn. Nonetheless, the negative starch/β-carotene correlation and the yet undefined textural characteristics has limited the actual adoption of improved orange-fleshed varieties. Therefore, comprehensive analysis of the genetic architecture of the negative association between starch and β-carotene could aid in advances of breaking this linkage as an important objective of breeding programs targeting sweetpotato for food and nutritional security.

Marker technology, including the utilization of SNPs and the determination of allele dosage, can be inferred through polyploid genotype calling methods (Pereira et al. 2018; Zych et al. 2019). Understanding the distribution of dosage-dependent key genes and alleles associated with high β-carotene content in cultivated sweetpotato can facilitate more efficient improvement of the crop. Currently, breeders can employ marker-assisted selection (MAS). For instance, findings from a dosage study in maize revealed a consistent increase in the concentrations of lutein, zeaxanthin, β-cryptoxanthin, and total carotenoids with the addition of each dominant Y1 allele to the endosperm, with the highest concentration observed at three doses (Egesel et al. 2003). MAS during the seedling stage is not only cost-effective (Slater et al. 2013) but also presents an attractive option for addressing recessive alleles, such as combining homozygous zeaxanthin epoxidase (zep) with dominant β-carotene hydroxylase to produce orange-fleshed tubers with significant zeaxanthin content in tetraploid potato (Wolters et al. 2010).

With a hexaploid genome, the breeding of β-carotene in sweetpotato can be evaluated using more advanced technologies. Strategies for metabolic engineering of the carotenoid pathway to increase β-carotene or enhance total carotenoid accumulation have been successfully implemented in plants (Bhatia and Ye 2012; Giuliano

2014). Three distinct and complementary strategies have been employed to enhance β-carotene accumulation in plants: overexpression of biosynthetic gene(s) ("push"), blocking the α-carotene branch pathway that competes with β-carotene biosynthesis and/or inhibiting the conversion of β-carotene to downstream products ("block"), and creating a sink for β-carotene accumulation by modulating the formation of chromoplasts or other carotenoid-sequestering structures ("sink"). Additionally, CRISPR/Cas-based gene editing can be utilized to fix desirable allelic variants, generate novel alleles, disrupt deleterious genetic linkages such as the beta carotene and starch, support pre-breeding efforts, and facilitate the introgression of favorable loci into elite lines.

Acknowledgements The authors Dr. Mercy Kitavi and Dr. C. Robin Buell are highly thankful for the financial support by a grant from the Bill and Melinda Gates Foundation.

References

Arscott SA, Tanumihardjo SA (2010) Carrots of many colors provide basic nutrition and bioavailable phytochemicals acting as a functional food. Compr Rev Food Sci Food Saf 9:223–239

Auldridge ME, Block A, Vogel JT, Dabney-Smith C, Mila I, Bouzayen M, Magallanes-Lundback M, DellaPenna D, McCarty DR, Klee HJ (2006) Characterization of three members of the *Arabidopsis carotenoid* cleavage dioxygenase family demonstrates the divergent roles of this multifunctional enzyme family. Plant J 45:982–993

Austin DF, Staples GW, Simao-Bianchini R (2015) A synopsis of Ipomoea (Convolvulaceae) in the Americas: further corrections, changes, and additions. Taxon 64:625–633

Bai C, Capell T, Berman J, Medina V, Sandmann G, Christou P, Zhu C (2016) Bottlenecks in carotenoid biosynthesis and accumulation in rice endosperm are influenced by the precursor-product balance. Plant Biotechnol J 14:195–205

Baranska M, Baranski R, Schulz H, Nothnagel T (2006) Tissue-specific accumulation of carotenoids in carrot roots. Planta 224:1028–1037

Berman J, Zorrilla-López U, Medina V, Farré G, Sandmann G, Capell T, Christou P, Zhu C (2017) The arabidopsis ORANGE (AtOR) gene promotes carotenoid accumulation in transgenic corn hybrids derived from parental lines with limited carotenoid pools. Plant Cell Rep 36:933–945

Beyer P, Al-Babili S, Ye X, Lucca P, Schaub P, Welsch R, Potrykus I (2002) Golden rice: introducing the β-carotene biosynthesis pathway into rice endosperm by genetic engineering to defeat vitamin A deficiency. J Nutr 132:506S-510S

Bhatia SK, Ye VM (2012) Metabolic engineering strategies for the production of beneficial carotenoids in plants. Food Sci Biotechnol 21:1511–1517

Bramley PM (2002) Regulation of carotenoid formation during tomato fruit ripening and development. J Exp Bot 53:2107–2113

Bridgers EN, Chinn MS, Truong V-D (2010) Extraction of anthocyanins from industrial purple-fleshed sweetpotatoes and enzymatic hydrolysis of residues for fermentable sugars. Ind Crops Prod 32:613–620

Britton G (1995) Structure and properties of carotenoids in relation to function. FASEB J 9:1551–1558

Burton GW, Ingold KU (1984) Beta-Carotene: an unusual type of lipid antioxidant. Science 224:569–573

Cao H, Wang J, Dong X, Han Y, Ma Q, Ding Y, Zhao F, Zhang J, Chen H, Xu Q, Xu J, Deng X (2015) Carotenoid accumulation affects redox status, starch metabolism, and flavonoid/anthocyanin accumulation in citrus. BMC Plant Biol 15:27

Carvalho LJ, Agustini MA, Anderson JV, Vieira EA, de Souza CR, Chen S, Schaal BA, Silva JP (2016) Natural variation in expression of genes associated with carotenoid biosynthesis and accumulation in cassava (*Manihot esculenta* Crantz) storage root. BMC Plant Biol 16:133

Cazzonelli CI, Pogson BJ (2010) Source to sink: regulation of carotenoid biosynthesis in plants. Trends Plant Sci 15:266–274

Chayut N et al (2017) Distinct mechanisms of the ORANGE protein in controlling carotenoid flux. Plant Physiol 173:376–389

Chen Y, Li F, Wurtzel ET (2010) Isolation and characterization of the Z-ISO gene encoding a missing component of carotenoid biosynthesis in plants. Plant Physiol 153:66–79

CIP (2020) CIP annual report 2019 discovery to impact: science-based solutions for global challenges (pyrmont, Australia: CIP), p 9

da Silva Pereira G, Gemenet DC, Mollinari M, Olukolu BA, Wood JC, Diaz F, Mosquera V, Gruneberg WJ, Khan A, Robin Buell C, Craig Yencho G, Zeng Z-B (2020) Multiple QTL mapping in autopolyploids: a random-effect model approach with application in a hexaploid sweetpotato full-sib population. Genetics 215:579–595

Dako E, Retta N, Desse G (2016) Comparison of three sweet potato (*I. batatas* (L.) Lam) varieties on nutritional and anti-nutritional factors. Global J Sci Front Res D Agricult Veter 16:1–11

Davison PA, Hunter CN, Horton P (2002) Overexpression of β-carotene hydroxylase enhances stress tolerance in Arabidopsis. Nature 418:203–206

Drapal M, Gerrish C, Fraser PD (2022) Changes in carbon allocation subplastidal amyloplast structures of specialised *I. batatas* (sweet potato) storage root phenotypes. Phytochemistry 203:113409

Egesel, CO, Wong, JC, Lambert, R, Rocheford, T (2003) Gene dosage effects on carotenoid concentration in maize grain

Eggersdorfer M, Wyss A (2018) Carotenoids in human nutrition and health. Arch Biochem Biophys 652:18–26

El-Sheikha AF, Ray RC (2017) Potential impacts of bioprocessing of sweet potato: review. Crit Rev Food Sci Nutr 57:455–471

FAOSTAT (2022) Production; Cassava, sweet potato, yams, taro; world; 1961–2019 (Online) Food and Agriculture Organization of the United Nations Downloaded data

Fernandez-Orozco R, Gallardo-Guerrero L, Hornero-Méndez D (2013) Carotenoid profiling in tubers of different potato (*Solanum* sp.) cultivars: accumulation of carotenoids mediated by xanthophyll esterification. Food Chem 141:2864–2872

Fiedor J, Burda K (2014) Potential role of carotenoids as antioxidants in human health and disease. Nutrients 6:466–488

Fraser PD, Bramley PM (2004) The biosynthesis and nutritional uses of carotenoids. Prog Lipid Res 43:228–265

Gemenet DC et al (2020) Quantitative trait loci and differential gene expression analyses reveal the genetic basis for negatively associated β-carotene and starch content in hexaploid sweetpotato [*I. batatas* (L.) Lam]. Theor Appl Genet 133:23–36

Giuliano G (2014) Plant carotenoids: genomics meets multi-gene engineering. Curr Opin Plant Biol 19:111–117

Giuliano G (2017) Provitamin A biofortification of crop plants: a gold rush with many miners. Curr Opin Biotechnol 44:169–180

Goodwin TW (1984) The biochemistry of the carotenoids. Springer, Netherlands

Grace MH, Yousef GG, Gustafson SJ, Truong V-D, Yencho GC, Lila MA (2014) Phytochemical changes in phenolics, anthocyanins, ascorbic acid, and carotenoids associated with sweetpotato storage and impacts on bioactive properties. Food Chem 145:717–724

Grune T, Lietz G, Palou A, Ross AC, Stahl W, Tang G, Thurnham D, Yin S-A, Biesalski HK (2010) Beta-carotene is an important vitamin A source for humans. J Nutr 140:2268S-2285S

Gurmu F, Hussein S, Laing M (2014) The potential of orange-fleshed sweet potato to prevent vitamin A deficiency in Africa. Int J Vitam Nutr Res 84:65–78

Han X, Ding S, Lu J, Li Y (2022) Global, regional, and national burdens of common micronutrient deficiencies from 1990 to 2019: a secondary trend analysis based on the global burden of disease 2019 study. eClinicalMedicine 44:11542

He W, Zeng M, Chen J, Jiao Y, Niu F, Tao G, Zhang S, Qin F, He Z (2016) Identification and quantitation of anthocyanins in purple-fleshed sweet potatoes cultivated in China by UPLC-PDA and UPLC-QTOF-MS/MS. J Agric Food Chem 64:171–177

Hirakawa H et al (2015) Survey of genome sequences in a wild sweet potato, *Ipomoea trifida* (H B K) G Don. DNA Res 22:171–179

Hirschberg J (2001) Carotenoid biosynthesis in flowering plants. Curr Opin Plant Biol 4:210–218

Horner HT, Healy RA, Ren G, Fritz D, Klyne A, Seames C, Thornburg RW (2007) Amyloplast to chromoplast conversion in developing ornamental tobacco floral nectaries provides sugar for nectar and antioxidants for protection. Am J Bot 94:12–24

Islam SN, Nusrat T, Begum P, Ahsan M (2016) Carotenoids and β-carotene in orange fleshed sweet potato: a possible solution to vitamin A deficiency. Food Chem 199:628–631

Jansen G, Flamme W (2006) Coloured potatoes (*Solanum tuberosum* L.): anthocyanin content and tuber quality. Genet Resour Crop Evol 53:1321–1331

Jaramillo AM, Sierra S, Chavarriaga-Aguirre P, Castillo DK, Gkanogiannis A, López-Lavalle LAB, Arciniegas JP, Sun T, Li L, Welsch R, Boy E, Álvarez D (2022) Characterization of cassava ORANGE proteins and their capability to increase provitamin A carotenoids accumulation. PLoS ONE 17:e0262412

Jarvis P, López-Juez E (2013) Biogenesis and homeostasis of chloroplasts and other plastids. Nat Rev Mol Cell Biol 14:787–802

Jenkins M, Shanks CB, Brouwer R, Houghtaling B (2018) Factors affecting farmers' willingness and ability to adopt and retain vitamin A-rich varieties of orange-fleshed sweet potato in Mozambique. Food Sec 10:1501–1519

Ji H, Zhang H, Li H, Li Y (2015) Analysis on the nutrition composition and antioxidant activity of different types of sweet potato cultivars. Food Nutr Sci 06:161–167

Kang L, Park S-C, Ji CY, Kim HS, Lee H-S, Kwak S-S (2017) Metabolic engineering of carotenoids in transgenic sweetpotato. Breed Sci 67:27–34

Kim JE, Rensing KH, Douglas CJ, Cheng KM (2010) Chromoplasts ultrastructure and estimated carotene content in root secondary phloem of different carrot varieties. Planta 231:549–558

Kim SH, Ahn YO, Ahn M-J, Jeong JC, Lee H-S, Kwak S-S (2013) Cloning and characterization of an Orange gene that increases carotenoid accumulation and salt stress tolerance in transgenic sweetpotato cultures. Plant Physiol Biochem 70:445–454

Kitahara K, Nakamura Y, Otani M, Hamada T, Nakayachi O, Takahata Y (2017) Carbohydrate components in sweetpotato storage roots: their diversities and genetic improvement. Breed Sci 67:62–72

Krinsky NI, Johnson EJ (2005) Carotenoid actions and their relation to health and disease. Mol Aspects Med 26:459–516

Kurnianingsih N, Ratnawati R, Nazwar TA, Ali M, Fatchiyah F (2020) A comparative study on nutritional value of purple sweet potatoes from West Java and Central Java. Indonesia. J Phys Conf Ser 1665:012011

Lee MJ, Park JS, Choi DS, Jung MY (2013) Characterization and quantitation of anthocyanins in purple-fleshed sweet potatoes cultivated in Korea by HPLC-DAD and HPLC-ESI-QTOF-MS/MS. J Agric Food Chem 61:3148–3158

Li L, Yuan H (2013) Chromoplast biogenesis and carotenoid accumulation. Arch Biochem Biophys 539:102–109

Li L, Yuan H, Zeng Y, Xu Q (2016) Plastids and carotenoid accumulation. Subcell Biochem 79:273–293

Liao Z, Chen M, Yang Y, Yang C, Fu Y, Zhang Q, Wang Q (2008) A new isopentenyl diphosphate isomerase gene from sweet potato: cloning, characterization color complementation. Biologia 63:221–226

Low JW, Thiele G (2020) Understanding innovation: the development and scaling of orange-fleshed sweetpotato in major African food systems. Agric Syst 179:102770

Low JW, Arimond M, Osman N, Cunguara B, Zano F, Tschirley D (2007) A food-based approach introducing orange-fleshed sweet potatoes increased vitamin A intake and serum retinol concentrations in young children in rural Mozambique. J Nutr 137:1320–1327

Low J, Lynam J, Lemaga B, Crissman C, Barker I, Thiele G, Namanda S, Wheatley C, Andrade M (2009) Sweetpotato in Sub-Saharan Africa. In: Loebenstein G, Thottappilly G (eds) The sweetpotato. Springer, Dordrecht, pp 359–390

Low JW, Mwanga ROM, Andrade M, Carey E, Ball A-M (2017) Tackling vitamin A deficiency with biofortified sweetpotato in sub-Saharan Africa. Global Food Sec 14:23–30

Lu S et al (2006) The Cauliflower or gene encodes a DnaJ cysteine-rich domain-containing protein that mediates high levels of β-carotene accumulation. Plant Cell 18:3594–3605

Lv Z, Yu K, Jin S, Ke W, Fei C, Cui P, Lu G (2019) Starch granules size distribution of sweet potato and their relationship with quality of dried and fried products. Starke 12:1800175

Mollinari, M, Olukolu, BA, da S Pereira, G, Khan, A, Gemenet, D, Craig Yencho, G, Zeng, Z-B (2020) Unraveling the hexaploid sweetpotato inheritance using ultra-dense multilocus mapping. G3 Genes Genom Genet 10:281–292

Moran NA, Jarvik T (2010) Lateral transfer of genes from fungi underlies carotenoid production in aphids. Science 328:624–627

Mortimer CL, Misawa N, Ducreux L, Campbell R, Bramley PM, Taylor M, Fraser PD (2016) Product stability sequestration mechanisms in *Solanum tuberosum* engineered to biosynthesize high value ketocarotenoids. Plant Biotechnol J 14:140–152

Mu J, Xu J, Wang L, Chen C, Chen P (2021) Anti-inflammatory effects of purple sweet potato anthocyanin extract in DSS-induced colitis: modulation of commensal bacteria attenuated bacterial intestinal infection. Food Funct 12:11503–11514

Neela S, Fanta SW (2019) Review on nutritional composition of orange-fleshed sweet potato its role in management of vitamin A deficiency. Food Sci Nutr 7:1920–1945

Niringiye CS, Ssemakula GN, Namakula J, Kigozi CB, Alajo A, Mpembe I, Mwanga ROM (2014) Evaluation of promising sweet potato clones in selected agro ecological zones of Uga Time. J Agricult Veter Sci

Odake K, Hatanaka A, Kajiwara T, Muroi T, Nishiyama K, Yamakawa O, Terahara N, Yamaguchi M (1994) Evaluation method and breeding of purple sweet potato "YAMAGAWA MURASAKI" (*I. batatas* POIR) for raw material of food colorants. Nippon Shokuhin Kogyo Gakkaishi 41:287–293

Olayide P, Large A, Stridh L, Rabbi I, Baldermann S, Stavolone L, Alexandersson E (2020) Gene expression and metabolite profiling of thirteen Nigerian cassava landraces to elucidate starch and carotenoid composition. Agronomy 10:424

Park S-C, Kim SH, Park S, Lee H-U, Lee JS, Park WS, Ahn M-J, Kim Y-H, Jeong JC, Lee H-S, Kwak S-S (2015) Enhanced accumulation of carotenoids in sweetpotato plants overexpressing IbOr-Ins gene in purple-fleshed sweetpotato cultivar. Plant Physiol Biochem 86:82–90

Pereira GS, Garcia AAF, Margarido GRA (2018) A fully automated pipeline for quantitative genotype calling from next generation sequencing data in autopolyploids. BMC Bioinform 19:398

Pogson BJ, Niyogi KK, Björkman O, DellaPenna D (1998) Altered xanthophyll compositions adversely affect chlorophyll accumulation and nonphotochemical quenching in *Arabidopsis* mutants. Proceed Natl Acad Sci 95:13324–13329

Rabbi IY, Udoh LI, Wolfe M, Parkes EY, Gedil MA, Dixon A, Ramu P, Jannink J-L, Kulakow P (2017) Genome-wide association mapping of correlated traits in cassava: dry matter and total carotenoid content. Plant Genome 10:134

Reeve RM (1967) A review of cellular structure, starch, and texture qualities of processed potatoes. Econ Bot 21:294–308

Ruiz-Sola MÁ, Rodríguez-Concepción M (2012) Carotenoid biosynthesis in Arabidopsis: a colorful pathway Arabidopsis. Book 10:e0158

Sandmann G, Römer S, Fraser PD (2006) Understanding carotenoid metabolism as a necessity for genetic engineering of crop plants. Metab Eng 8:291–302

Sauer L, Li B, Bernstein PS (2019) Ocular carotenoid status in health and disease. Annu Rev Nutr 39:95–120

Scott GJ, Ewell PT (1992) Sweetpotato in African food systems. Product Develop Root Tuber Crops 3:91–103

Slater AT, Cogan NOI, Forster JW (2013) Cost analysis of the application of marker-assisted selection in potato breeding. Mol Breed 32:299–310

Steed LE, Truong V-D (2008) Anthocyanin content, antioxidant activity, and selected physical properties of flowable purple-fleshed sweetpotato purees. J Food Sci 73:S215–S221

Sun T, Yuan H, Cao H, Yazdani M, Tadmor Y, Li L (2018) Carotenoid metabolism in plants: the role of plastids. Mol Plant 11:58–74

Tzuri G et al (2015) A "golden" SNP in CmOr governs the fruit flesh color of melon (*Cucumis melo*). Plant J 82:267–279

Vallabhaneni R, Wurtzel ET (2009) Timing biosynthetic potential for carotenoid accumulation in genetically diverse germplasm of maize. Plant Physiol 150:562–572

van Jaarsveld PJ, Faber M, Tanumihardjo SA, Nestel P, Lombard CJ, Benadé AJS (2005) β-Carotene–rich orange-fleshed sweet potato improves the vitamin A status of primary school children assessed with the modified-relative-dose-response test. Am J Clin Nutr 81:1080–1087

Vimala B, Nambisan B, Hariprakash B (2011) Retention of carotenoids in orange-fleshed sweet potato during processing. J Food Sci Technol 48:520–524

von Lintig J (2012) Provitamin A metabolism and functions in mammalian biology. Am J Clin Nutr 96:1234S-S1244

Wadl PA, Olukolu BA, Branham SE, Jarret RL, Yencho GC, Jackson DM (2018) Genetic diversity and population structure of the USDA sweetpotato (*I. batatas*) germplasm collections using GBSpoly. Front Plant Sci 9:1166

Wang A, Li R, Ren L, Gao X, Zhang Y, Ma Z, Ma D, Luo Y (2018) A comparative metabolomics study of flavonoids in sweet potato with different flesh colors (*I. batatas* (L.) Lam). Food Chem 260:124–134

Wolters A-MA, Uitdewilligen JGAML, Kloosterman BA, Hutten RCB, Visser RGF, van Eck HJ (2010) Identification of alleles of carotenoid pathway genes important for zeaxanthin accumulation in potato tubers. Plant Mol Biol 73:659–671

Woolfe JA (1992) Sweet potato an untapped food

World Food Prize Foundation (2016) https://www.worldfoodprize.org/index.cfm/87428/40771/2016_world_food_prize_presented_to_four_international_researchers

Wu S et al (2018) Genome sequences of two diploid wild relatives of cultivated sweetpotato reveal targets for genetic improvement. Nat Commun 9:4580

Wurtzel ET (2019) Changing form and function through carotenoids and synthetic. Biol Plant Physiol 179:830–843

Wurtzel ET, Cuttriss A, Vallabhaneni R (2012) Maize provitamin a carotenoids, current resources, and future metabolic engineering challenges. Front Plant Sci 3:29

Xu, J, Su, X, Lim, S, Griffin, J, Carey, E, Katz, B, Tomich, J, Smith, JS, Wang, W (2015) Characterisation and stability of anthocyanins in purple-fleshed sweet potato. P40 Food Chem 186:90–96

Yada B, Brown-Guedira G, Alajo A, Ssemakula GN, Owusu-Mensah E, Carey EE, Mwanga ROM, Yencho GC (2017) Genetic analysis and association of simple sequence repeat markers with storage root yield, dry matter, starch and β-carotene content in sweetpotato. Breed Sci 67:140–150

Yazdani M et al (2019) Ectopic expression of ORANGE promotes carotenoid accumulation and fruit development in tomato. Plant Biotechnol J 17:33–49

Yoshinaga M, Yamakawa O, Nakatani M (1999) Genotypic diversity of anthocyanin content composition in purple-fleshed sweet potato (*I. batatas* (L.) Lam). Breed Sci 49:43–47

Zakar T, Laczko-Dobos H, Toth TN, Gombos Z (2016) Carotenoids assist in cyanobacterial photosystem II assembly and function. Front Plant Sci 7:295

Zhang MK, Zhang MP, Mazourek M, Tadmor Y, Li L (2014) Regulatory control of carotenoid accumulation in winter squash during storage. Planta 240:1063–1074

Zhou X, Welsch R, Yang Y, Álvarez D, Riediger M, Yuan H, Fish T, Liu J, Thannhauser TW, Li L (2015) *Arabidopsis* OR proteins are the major posttranscriptional regulators of phytoene synthase in controlling carotenoid biosynthesis. Proceed Natl Acad Sci 112:3558–3563

Zych K, Gort G, Maliepaard CA, Jansen RC, Voorrips RE (2019) FitTetra 20: improved genotype calling for tetraploids with multiple population and parental data support. BMC Bioinform 20:148

Advances in Our Understanding of the Genetic Regulation of Storage Root Formation and Growth

9

Arthur Villordon and Don LaBonte

Abstract

Storage root formation is the most economically important developmental process in sweetpotato. Despite recent progress in the physiological and molecular understanding of how storage roots form, significant knowledge gaps exist in terms of explaining the variable number of storage roots produced per plant. Does the onset of storage root formation occur at random times in random adventitious roots, or is this process initiated by spatial and temporal cues in the rhizosphere that interact with shoot-borne signals? This review addresses this question and focuses on the vascular cambium as the main driver of storage root formation, which is essentially secondary growth. The goal is to integrate classical source-sink dynamics with available anatomical, morphological, physiological, molecular, and genomic evidence, leading to a more complete understanding of the genetic regulation of the role of vascular cambium in sweetpotato storage root formation. The understanding of how adventitious roots transition to storage roots is important not only from the scientific understanding but can lead to practical applications that improve food security and economic sustainability where the sweetpotato is grown.

Keywords

Adventitious roots · Vascular cambium · Root architecture · Sink strength

9.1 Introduction

Sweetpotato [*Ipomoea batatas* (L.) Lam.] is recognized as the seventh most important food crop in the world (FAOSTAT data 2019). It has a global production of approximately 144 million metric tons and is the third most important root or tuber crop after potato (*Solanum tuberosum* L.) and cassava (*Manihot esculenta* Crantz) (FAOSTAT 2019). World production is centered in the Asian-Southeast Asian region, with China being the largest producer, while Sub-Saharan Africa ranks second (FAOSTAT 2019). Although leaves are consumed as vegetables in some regions, the fleshy storage roots are the main economically important organ of the crop that is grown in diverse production

A. Villordon (✉)
LSU AgCenter Sweet Potato Research Station, 130 Sweet Potato Road, Chase, LA 71324, USA
e-mail: avillordon@agcenter.lsu.edu

D. LaBonte
School of Plant, Environmental and Soil Sciences, 131 J.C. Miller Hall, Louisiana State University, Baton Rouge, LA 70803, USA
e-mail: dlabonte@agctr.lsu.edu

© The Author(s) 2025
G. C. Yencho et al. (eds.), *The Sweetpotato Genome*, Compendium of Plant Genomes,
https://doi.org/10.1007/978-3-031-65003-1_9

environments with yields ranging from 4–10 t/ha in SSA (Ngailo et al. 2019) to 60–90 t/ha in the sweetpotato growing regions of Australia (Stirling 2021). Sweetpotato is used as a source of starch, ethanol, and animal fodder in most of Asia while it is considered a subsistence crop in Africa. The USA, Israel, Japan, Australia, New Zealand, and South Africa are among the few countries that grow sweetpotato as a vegetable to market in developed economies. In many production environments, the storage root is also the basis of propagation. Depending on the number of adventitious roots that will be induced to form storage roots, sweetpotato plants will yield either a high number (4–8/plant) of marketable storage roots or a low number of roots that may even be reduced to one very large storage root per plant or no marketable roots at all. Poor shape is another quality variable that renders many roots unmarketable. Due to the underground nature of the crop, the performance of sweetpotato plants can be evaluated only post-factum and the above-ground growth provides little or no indication of crop yield during development. Substantial information is missing with relation to the formation of sweetpotato storage roots in general. There have been recent reviews (Ravi et al. 2009; Tanaka 2016; Yang et al. 2023) of the progress of physiological and molecular mechanisms of storage root formation in sweetpotato. While available anatomical and molecular evidence has significantly increased our understanding of how storage roots form, significant knowledge gaps exist in terms of explaining the variable number of storage roots produced per plant. Thus, the understanding of how adventitious roots transition to storage roots is important not only from the scientific understanding but can lead to practical applications that improve food security and economic sustainability where the sweetpotato is grown.

Hoang et al. (2020a, b) took a cogent approach in synthesizing available evidence on storage root development in root crops by focusing on the vascular cambium as the main driver of secondary growth. Hoang et al. (2020a, b) concluded that the amount and resolution of currently available data for each root crop create gaps as far as pinpointing the key processes responsible for crop yield. This lack of granularity leads to a lack of resolution as regards tissue-specific gene expression patterns despite hundreds of thousands of candidate genes identified. This review coincided with the publication by Hoang et al. (2020a, b) of their work on identifying conserved gene regulatory networks of secondary growth in radish. In this work, they used laser capture microdissection to collect tissue samples for gene expression analysis from multiple time points representing key development stages of radish storage roots. They then compared to putative *Arabidopsis* orthologues to gain insights about gene regulation pathways in radish. Blomster and Mähönen (2020) reviewed this work and suggested that this tissue-specific radish dataset may well help to elucidate additional regulators radial growth and that the radish storage root could serve as an informative model for storage organ development. Blomster and Mähönen (2020) highlighted the underlying challenge in elucidating the genetic regulation of storage roots of root crops in general and the sweetpotato in particular: the lack of a model system.

This review will highlight studies that focus on vascular cambium as the main driver of storage root formation in sweetpotato. The overarching goal of this review is to integrate classical anatomical benchmarks, physiological studies, and emerging root architecture evidence into recent genetic studies to provide an updated summary of the genetic regulation of storage root formation. Focusing on the cambium enables the integration of various lines of evidence, including the role of hormone signaling. This review will also address carbon partitioning and differential sink strength to explain why under certain conditions, some adventitious roots fail to become storage roots. Further storage root growth is due to the translocation of photosynthates produced in the leaves to the developing storage roots where the sucrose is converted to other forms of carbohydrates. This review will also address the inconsistencies in terminology, specifically as it applies to the root system, as used in the majority of past and recent work that seek to elucidate the genetic regulation of storage root formation and development in sweetpotato. The use of inconsistent root terminology in root biology hinders understanding and scientific progress (Dubrovsky 2022; Zobel and Waisel 2010). Finally, we will present a model of storage root formation that is based on current available evidence. This model will address current gaps in knowledge, in particular offering proposed mechanisms that lead to the failure of some adventitious roots to become storage roots in response to environmental and management variables.

9.2 Anatomical Benchmarks

Classical and recent morphological and anatomical studies have unambiguously defined the anatomical features associated with the transition from primary to secondary growth associated with storage root formation in sweetpotato. Foremost of this is the work of McCormick (1916) that not only set the tone for future work but clearly documented the secondary features associated with storage root thickening, noting the role of primary and secondary cambium (Fig. 9.1c, d). McCormick also noted that there were as many rows of lateral roots as protoxylem points and explains the presence of definite

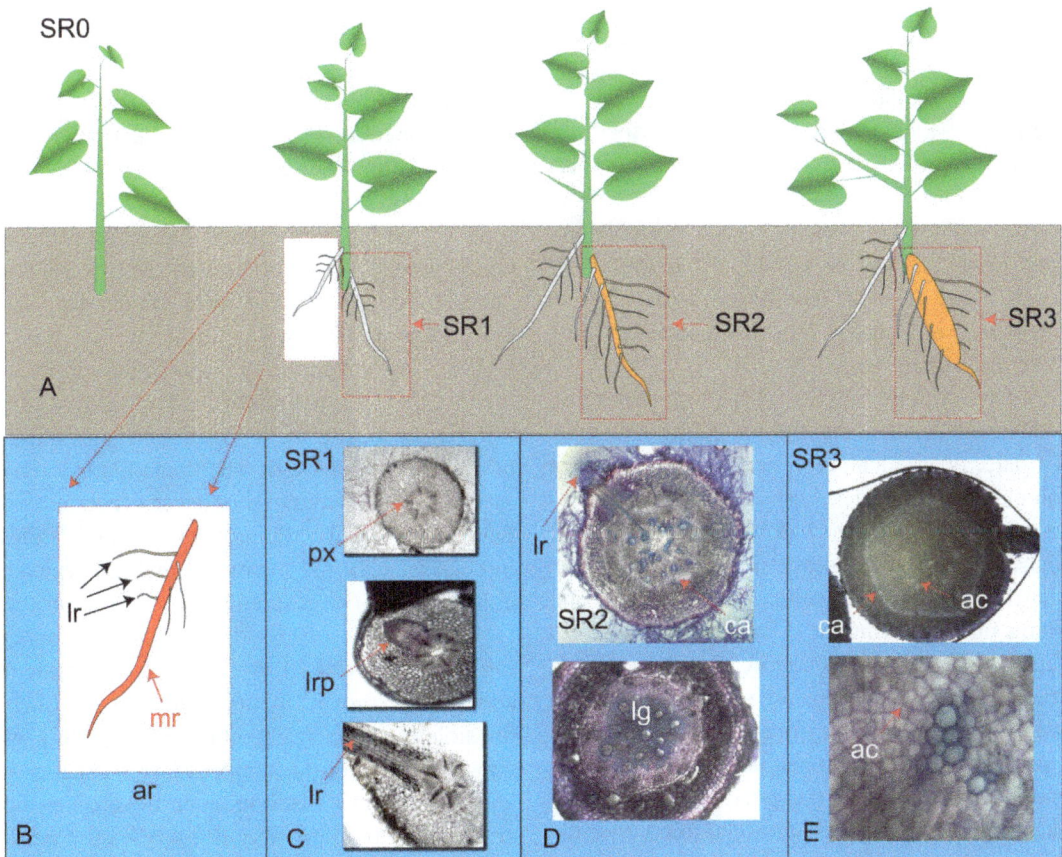

Fig. 9.1 SR1—primary growth, differentiation of protoxylem (5 DAP). SR2—Onset of secondary growth marked by the appearance of vascular cambium (ca) (15–20 DAP). SR3—appearance of anomalous cambium (20–25 DAP)

rows of lateral roots on "mature" storage roots (Fig. 9.1b). Artschwager (1924) corroborated the association of primary and secondary cambium with the early differentiation of the fleshy root. Reference was also made about the presence of lateral roots which sit in a scar-like tissue similar to the "potato eye." In addition, Artschwager (1924) noted that the "parts of the sweetpotato roots that do not become thickened," the cells between the protoxylem points and the large central cell become lignified. At this point, anatomical studies were largely descriptive. Togari (1950) built on these prior works and documented the roles of nutrients, temperature, water, and light on anatomical cues of storage root development. More importantly, Togari (1950) established the timing of anatomical benchmarks that were associated with storage root formation (Fig. 9.1). Defining these stages was important as it provided a context for gene expression analysis. Esau (1967) synthesized available evidence and described the secondary growth in sweetpotato as a complex type of anomalous growth in fleshy adventitious roots. Esau (1967) used the terms "normal cambium" and "anomalous cambium." In this work, we will use "primary cambium" and "vascular cambium" interchangeably, to distinguish from "anomalous cambium."

Defining these stages is important as it provides context for unraveling signaling networks. To date, anatomical evidence of anomalous or circular cambium development remains the key indicator of the onset of storage root formation. Gene expression studies by nature are time-sensitive and tissue-specific assays and the outcome may be different if sampled at different time points. When Firon et al. (2013) generated transcriptome data from adventitious roots that were either undergoing storage root formation or lignification, they sectioned adventitious roots at the 2.5 cm section of the proximal tip, verified the anatomical features, and classified root tissue samples accordingly.

9.3 Emerging Root Architecture Terminology: Consensus or Conundrum?

Gregory and Wojciechowski (2020) conducted a comprehensive review of the literature of root systems of root and tuber crops and noted that the inconsistency in terminology applied to root systems of these crops was a notable feature of their effort to synthesize the available literature. In particular, they highlighted the incorrect application of terms used to describe the root systems which, with the exception of a few cassava and yam crops grown from seed, all of the root and tuber crops produce adventitious roots (ARs) (Fig. 9.1a, b). They also agreed with prior work (Adu et al. 2018; Villordon et al. 2014) that the term "fibrous roots" is unhelpful and misleading. AR axes (the main root) emerge from stem nodes, basal stems of cells (wound tissues), stolons, and the junction of stem and mother tuber/corm of the crop and lateral roots emerge from these axes. Esau (1967) defined "lateral root" as any root branching from another root. To assist in describing the relationship of lateral roots, lateral root orders are described as "first-order laterals" and from these arise "second-order laterals" and so on (Zobel and Waisel 2010). This distinction is important because lateral roots are functionally and physiologically different from the main axis (or the main root) of the adventitious root where cambium activity associated with storage root development occurs. Lateral roots have diarch or polyarch steles, in contrast with the main root that are typically either pentarch or hexarch steles.

Why is the term "fibrous root" confusing in the context of sweetpotato root systems and storage root formation? In a review of the physiology of the sweetpotato, Kays (1985) discussed separately the subject of root distribution and architecture from sections devoted to lateral roots, "primary fibrous roots," and "pencil roots." Kays (1985) described "primary fibrous roots" as emerging largely from tetrarch "thin" adventitious roots although "under adverse conditions, they maybe from pentarch, hexarch, and even septarch thick roots." Earlier, Wilson and Lowe (1973) reported on the anatomical features of field-grown sweetpotato, corroborated the timing of anatomical benchmarks proposed by Togari (1950), and introduced some variation in terminology. For example, they referred to "tuberous" and "non-tuberous" roots and introduced the term "fibrous roots," which they defined as uniformly thickened roots with normal secondary growth leading to complete lignification of the stele. In other words, Wilson and Lowe (1973) used the term "fibrous roots" in reference to non-swollen ARs. Belehu et al. (2004) determined that "fibrous roots" were lateral roots and proposed the term to refer to first, second, and third-order lateral roots.

9.4 Sink Strength: Bridges Gap Between Morpho-Anatomical and Molecular Data?

Scientists have repeatedly recognized that species with large below-ground sinks for carbon and with apoplastic mechanisms of phloem loading are likely to be the best candidates for a large response to rising atmospheric CO_2 (Miglietta et al. 2000). There is also an increasing consensus that growth or storage sink limitations are possibly major factors constraining responses of plants to elevated CO_2 (Miglietta et al. 2000). To understand the problem of regulation of dry matter partitioning by the sinks, there has been substantial interest in a property of a sink, called sink strength, that determines this regulation. Sink strength can be defined as the competitive ability of an organ to receive or attract assimilates (Wareing and Patrick 1975; Wolswinkel 1985; Farrar 1993a). At present, many discussions focus on the question whether the concept of sink strength is a useful one, or a vague and confusing concept (Farrar 1993b). Much confusion is due to lack of a clear definition of sink strength. The actual rate of assimilated import or growth has often been used as a measure of sink

strength (Warren-Wilson 1972). When defined in this way, sink strength in fact represents the net result of assimilate flow which may depend on the competitive ability of all sinks on a plant and the assimilates supply (source strength). This is not a useful measure of sink strength, and it is the prime cause why some authors reject the use of the concept of sink strength. Minchin and Thorpe (1993), dismissed sink strength (as measured by the actual import rate) as a misnomer, and other authors (e.g., Patrick 1993) stated that it should be possible to identify a set of parameters to describe a sink's ability to influence assimilate import which are independent of the rest of the plant. More recent evidence in other species also supports the hypothesis that RSA contributes to the determination of sink strength and is consistent with increased upregulation of enzymes involved in starch and sucrose metabolism (reviewed by Hennion et al. 2019). Current available evidence about the role of lateral root emergence and RSA in determining sink strength is consistent with modeling work (Bidel et al. 2000; Thaler and Pages 1998). Linking sucrose synthase (SuSy) activity with RSA in sweetpotato fills significant gaps in our knowledge of storage root formation, strengthens the concept of sink strength, and can contribute to advances in water and nutrient management and contribute to harnessing high-value root traits for crop improvement. New evidence from RSA studies (Bui et al. 2015; Paszkowski and Gutjahr 2013) supports the hypothesis that root architecture is a key determinant of root carbon sink strength, integrating current available molecular, hormonal, nutritional, and morphological evidence that leads to a more comprehensive understanding of storage root formation.

9.5 Genetic Regulation of Vascular Cambium-Driven Storage Root Formation in Sweetpotato: A Synthesis

The overarching goal of this review is to highlight work that defines storage root formation within the context of vascular cambium development, with clear definition of developmental stages either via anatomical features, and well-described root developmental stages and root orders. For clarity, if the term "fibrous roots" is used other than in reference to "lateral roots," as defined previously, then it will be noted. This review will also highlight genetic data associated with lateral root development, protoxylem development, carbon allocation, and lignification presented within the context of storage root formation as defined in Fig. 9.1.

Table 9.1 highlights studies that specifically define storage root formation within the context of vascular cambium development and with the objective of identifying genes or genetic networks associated with storage root formation in sweetpotato. The work by You et al. (2003) likely represented the first attempt to identify genes associated with storage root formation in sweetpotato. They classified adventitious roots based on thickness and generated a cDNA library based on their definition of "early stage storage roots (0.3–1 cm in diameter)." They sequenced the clones and identified 39 genes putatively involved in gene regulation, signal transduction, and development. Of these 39 genes, *IbMADS3* and *IbMADS4* were categorically associated with cambium development.

Table 9.1 List of genes that have been associated with specific storage root formation stages and benchmarks in sweetpotato

Storage root formation stage				Storage root formation benchmark[a]				Reference
Protoxylem development; lateral root emergence;	Lignification	Vascular cambium and carbon allocation	Anomalous cambium	RO	OT	AC	TS	
			IbMADS3, IbMADS4	−	+	−	−	You et al. (2003)
		SRF6	*SRF6*	−	+	+	−	Tanaka et al. (2005)
	IbAGL17			−	−	+	−	Kim et al. (2005)
IbMADS1				−	+	+	−	Ku et al. (2008)
		SRD1	*SRD1*	−	+	+	−	Noh et al. (2010)
		Suy		−	+	−	−	Tao et al. (2012)
	4CL, CCoAOMT, CAD			−	+	+	−	Firon et al. (2013)
	IbEXP1			−	+	+	−	Noh et al. (2013)
				−	+	+	−	Wang et al. (2015)
		LBD4, WOX4, TMO6, GLGL, SSY, GLGB	*LBD4, WOX4, TMO6*	−	+	+	−	Dong et al. (2019)
IbKN2, IbKN3	*IbPAL, IbC4H, Ib4CL, IbCCoAOM, IbCAD*	*IbAGPAse, IbGBSS*		+	+	+	−	Singh et al. (2019)
		SuSy, SS, SPS, INV	*AUX/IAA, ARF, SAUR, CH3*	−	+	−	−	Cai et al. (2022)
		SBEI		−	+	−	−	Song et al. (2022)

[a] Benchmarks for elucidating genetic regulation of storage root formation as established by Hoang et al. (2020a, b) RO = root system and root order clearly defined as an adventitious root system and lateral roots arising from the main axis (or main root). If "fibrous roots" is used, it must include images or illustrations at various time points corresponding to the key stages defined in Fig. 9.1 (with "fibrous roots" clearly identified), along with confirmation of key anatomical features; OT = ontogeny and timing = time points sampling corresponding to storage root development as per Fig. 9.1 or at the minimum provisionally defined time points which may deviate from Fig. 9.1; AC = anatomical confirmation; TS = tissue specificity

Tanaka et al. (2005) used digoxigenin (DIG) labeling to specifically link *SRF6* expression to the vascular cambium and anomalous cambium (Table 9.1). This is likely the first evidence linked gene expression data to anatomical location. They also detected genes associated with sugar metabolism, signal transduction, and carotenoid biosynthesis. Kim et al. (2005) used a candidate gene approach to link *IbAGL17*, a MADS-box gene to increased sink strength of developing adventitious roots. It was also through DIG labeling that Ku et al. (2008) localized *IbMADS1* expression to within the stele and lateral root primordia, effectively linking lateral root emergence with storage root formation. In this work, Ku et al. (2008) used "fibrous roots" apparently to refer to adventitious roots in various stages of storage root formation.

However, the anatomical evidence clearly showed that the gene expression data was linked to lateral root emergence sites. Ku et al. (2008) concluded that *IbMADS1* is an important integrator at the initiation of storage root formation and possibly regulated by a network involving a MADS-box gene in which hormones such as jasmonic acid and cytokinins are trigger factors. Noh et al. (2010) presented DIG labeling evidence that *SRD1* expression was localized in cambial cells but no signal was detected in storage parenchyma and xylem vessels. They also presented evidence that SRD1 was responsive to variation in auxin concentration. The work by Tao et al. (2012) likely represented the first use of next-generation RNA sequencing that made possible the investigation of storage root development without genome sequence information. Even though the experimental methodology precluded the identification of genes associated with storage root initiation and early development, Tao et al. (2012) presented evidence of increased SuSy expression associated with expanding storage roots. Firon et al. (2013) likely represented the first work that coupled anatomical confirmation of anomalous cambium with NGS transcriptome data. More importantly, this work specified the specific section of the main adventitious root where the evidence of anomalous cambium was detected and corresponded to the tissue used for RNA extraction. This work provided evidence of upregulation of genes involved in carbohydrate and starch biosynthesis and the downregulation of genes (*4CL, CCoAOMT, CAD*) involved in lignin biosynthesis in adventitious roots that failed to show evidence of storage root formation. Noh et al. (2013) presented evidence that an expansin-like gene, *IbEXP1*, was apparently negatively involved in storage root formation by suppressing the proliferation of metaxylem and cambium cells. Wang et al. (2015) used microarray data to generate evidence that starch biosynthesis is upregulated while lignin biosynthesis is downregulated during storage root development. In addition, this work provided evidence that transcription factors that modulate or control root development and lateral root were also detected during storage root development. However, no specific genes were associated with vascular cambium development. Dong et al. (2019) analyzed transcriptome data from different developmental stages based on adventitious root diameter. They used "fibrous roots" to refer to adventitious roots less than 1 cm in diameter and apparently to describe adventitious roots that do not show evidence of vascular or anomalous cambium development. This work identified *LBD4, WOX4*, and *TMO6* were associated with cambium activity. It also identified starch biosynthesis genes, including ADP-glucose pyrophosphorylase (GLGL), starch synthase (SSY), and starch-branching enzyme (GLGB). Singh et al. (2019) focused on the role of GA on storage root formation and provided evidence that GA suppressed cambium development and was associated with lignin biosynthesis. This work also provided evidence that lateral root development was linked to the capacity of an adventitious root to undergo storage root formation. Singh et al. (2019) presented a model accounting for the role of GA in storage root formation that incorporates root architecture attributes like lateral root number and length. However, this model does not provide any links to external management or environmental variables known to suppress storage root formation or favor lignification. Cai et al. (2022)

also used adventitious root thickness to clas-sify storage root formation stages to character-ize genes associated with storage root formation using NGS. They identified auxin-responsive genes (AUX/IAA, ARF, SAUR, and CH3) that were associated with anomalous cambium activ-ity. Interestingly, they provided evidence that expansin was associated with storage root devel-opment, contrary to the evidence presented by Noh et al. (2013). An examination of the meth-odology used to generate tissue samples in both studies showed significant differences in terms of timing and specificity of tissue collection. Both studies used the term "fibrous roots" to refer to adventitious roots that are not associated with storage root formation. However, Noh et al. (2013) defined the diameter of "fibrous roots" as less than 0.2 cm while Cai et al. (2022), defined this as less than 0.1 cm. Prior work has indicated that adventitious roots around 1 mm can show evidence of storage root formation depending on cultivar and growth conditions (Wilson and Lowe 1973). Furthermore, Noh et al. (2013) did not define the sampling time while Cai et al. (2022) collected storage root samples at 90 days after planting. Both studies did not specify the specific tissue from which samples were col-lected for RNA extraction. This lack of standard experimental protocols and specificity of tissue sampling complicates the direct comparison of gene expression data. Cai et al. (2022) also developed a model for storage root formation based on their transcriptomic data, highlighting the role of genes hypothesized to be involved in storage root formation and outlining presump-tive regulatory pathways. As with the prior cited work, this hypothetical model does not provide alternate pathways for lignification nor proposes any links to known external variables that affect storage root formation like moisture and temperature. Both models highlight the role of starch and sucrose metabolism. The role of starch metabolism in storage root formation is further highlighted by Song et al. (2022) who performed comparative transcriptomics using NGS. They also identified transcription factors possibly associated with storage root forma-tion. They concluded that additional research is

needed in order to validate their roles in storage root formation.

Taken together, recent molecular work under-scores the role of starch biosynthesis genes in storage root formation. However, due in part to the lack of hypothetical models in stor-age root formation that account for variation in sink strength among adventitious roots, these genes are merely enumerated and not prop-erly contextualized. Ravi et al. (2009) stated that sink strength determined storage root growth but did not elaborate on how this varied among adventitious roots. Li and Zhang cor-related sink strength with greater SuSy in stor-age root expressed sequence tags (ESTs) than in non-storage root ESTs and correlated with ADP-glucose pyrophosphorylase (AGPase) expression. AGPAse catalyzes the formation of ADP-glucose, the first step dedicated to starch synthesis (reviewed in Hennion et al. 2019). In sweetpotato, cumulative evidence supports the hypothesis that sink strength determines the capacity of adventitious roots to undergo storage root formation (Keutgen et al. 2002; Li and Zhang 2003). Considering prior evi-dence, it is therefore surprising that the concept of sink strength has been overlooked in most of the reports cited in this review. Hoang et al. (2020a, b) and Zierer et al. (2021) reviewed the physiological and genetic regulation of storage roots in general and specifically addressed the role of sink strength in storage root formation. Prior work assumed that all adventitious roots were phenotypically uniform, and the subject of RSA variability was never accounted for in the context of storage root formation signaling. However, cumulative data from recent research indicate that adventitious roots within the same plant vary in root architectural attributes (main root length, lateral root number, lateral root length, lateral root density) in response to biotic and abiotic variables. These root architectural modifications in turn are associated with the competency of an adventitious root to undergo storage root formation. Evidence presented by Ku et al. (2008) and Singh et al. (2019) under-scores the importance of incorporating lateral root measurements in current and future work that seeks to further characterize the genetic regulation of storage root formation in sweetpotato.

A model depicting the synthesis and integration of available molecular evidence of vascular cambium-driven storage root formation into existing anatomical and physiological benchmarks is depicted in Fig. 9.2. Carbon partitioning and sink strength determination are also incorporated into the model, along with pathways hypothesizing the sensing of environmental cues. In this model, the developing root system integrates internal and external signals that in turn

determine sink strength of individual storage roots. Adventitious roots that develop into the soil profile characterized as possessing optimal conditions (temperature, moisture, fertility) will develop optimal lateral root architecture, which in turn increases sink strength. It is hypothesized that sucrose is a shoot-derived signal associated with vascular cambium development. On the other hand, adventitious roots that develop into marginal soil conditions fail to develop optimal root architecture, reducing its sink strength, and unable to compete for carbon allocation.

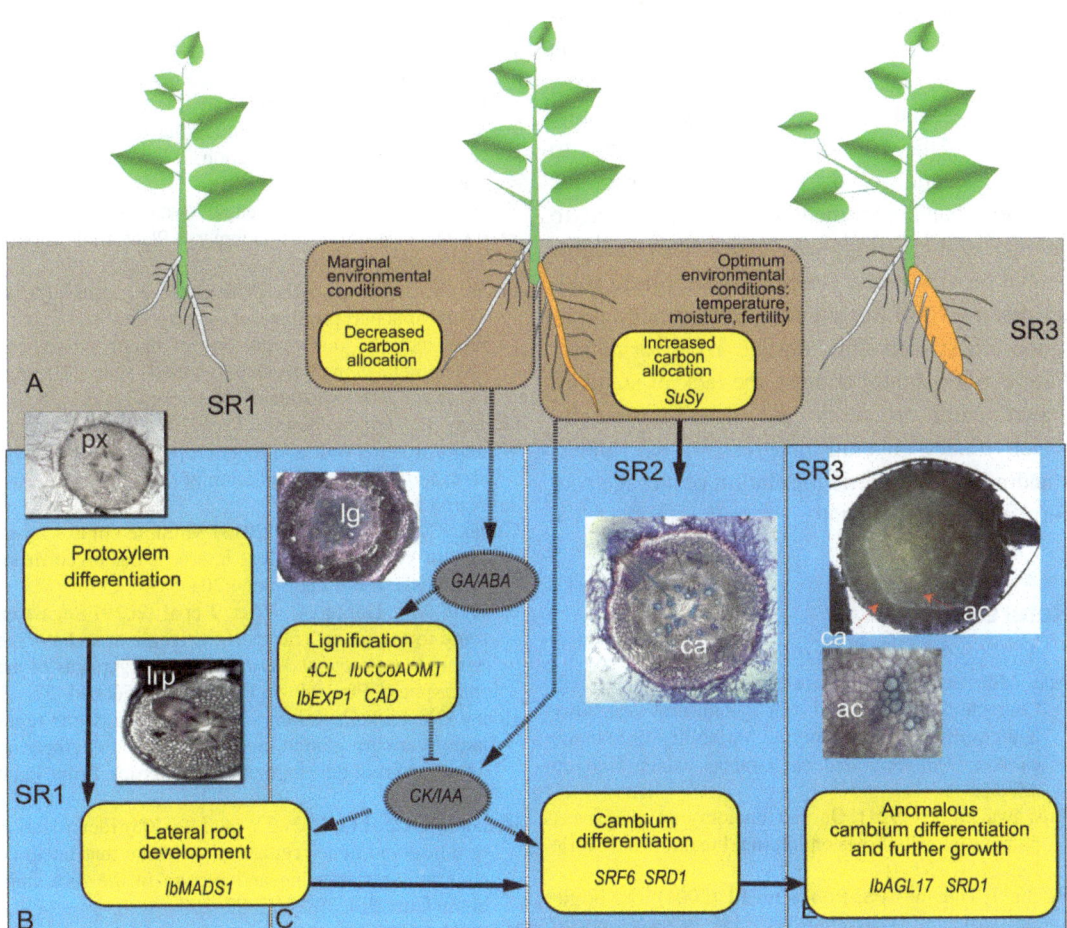

Fig. 9.2 Hypothetical model synthesizing current understanding of genetic regulation of storage root formation that integrates sink strength determination and root architectural responses to external management and environmental stimuli

9.6 Conclusions

It is evident from the current review that the integration of anatomical, morphological, and physiological cues of storage root formation with molecular and genomic evidence will lead to a more complete understanding of the genetic regulation of the role of vascular cambium in sweetpotato storage root formation. Recent technological advances and the availability of reference genomes have led to significant advances in unraveling the genetic regulation of storage root formation in sweetpotato. At the same time, increased attention to the role of vascular cambium in other root crops such as radish and cassava has underscored the need to improve the resolution and tissue specificity of gene expression studies and consolidation of genomic and molecular data. This consolidation will lead to the identification of shared regulatory programs and promote comprehensive studies related to storage root development. These new insights should provide a new benchmark for future studies that seek to further unravel the genetic regulation of sweetpotato storage root formation. Future work should address the lack of standard experimental protocols and tissue specificity which hinders overall progress in our understanding of the genetic regulation of storage root formation in sweetpotato.

References

Adu MO, Asare PA, Asare-Bediako E et al (2018) Characterising shoot and root system trait variability and contribution to genotypic variability in juvenile cassava (*Manihot esculenta* Crantz) plants. Heliyon 4:00665

Artschwager E (1924) On the anatomy of the sweet potato root, with notes on internal breakdown. J Agric Res 27(3):157–166

Belehu T, Hammes PS, Robbertse PJ (2004) The origin and structure of adventitious roots in sweet potato (Ipomoea batatas). Aust J Bot 52(4):551–558

Bidel LPR, Pagès L, Riviere LM et al (2000) MassFlowDyn I: a carbon transport and partitioning model for root system architecture. Ann Bot 85(6): 869–886

Blomster T, Mähönen AP (2020) Plant biology: storage root growth through thick and thin. Curr Biol 30(15):R880–R883

Bui HH, Serra V, Pagès L (2015) Root system development and architecture in various genotypes of the Solanaceae family. Botany 93(8):465–474

Cai Z, Cai Z, Huang J et al (2022) Transcriptomic analysis of tuberous root in two sweet potato varieties reveals the important genes and regulatory pathways in tuberous root development. BMC Genom 23(1):1–19

Dong T, Zhu M, Yu J et al (2019) RNA-Seq and iTRAQ reveal multiple pathways involved in storage root formation and development in sweet potato (*I. batatas* L.). BMC Plant Biol 19(1):1–16

Dubrovsky JG (2022) Inconsistencies in the root biology terminology: let's communicate better. Plant Soil 476(1):713–720

Esau K (1967) Plant anatomy. 2nd edn. Wiley, New York

FAO (2019) Food and Agriculture Organization of the United Nations. http://www.fao.org/faostat/en/#data/QC. Accessed 20 Feb 2024

Farrar JF (1993a) Sink strength: What is it and how do we measure it? Introduction. Plant Cell Environ 16(9):1015

Farrar JF (1993b) Sink strength: what is it and how do we measure it? A summary. Plant Cell Environ 16(9):1045–1046

Firon N, LaBonte D, Villordon A et al (2013) Transcriptional profiling of sweetpotato (*I. batatas*) roots indicates down-regulation of lignin biosynthesis and up-regulation of starch biosynthesis at an early stage of storage root formation. BMC Genom 14(1):1–25

Gregory PJ, Wojciechowski T (2020) Root systems of major tropical root and tuber crops: root architecture, size, and growth and initiation of storage organs. Adv Agron 161:1–25

Gutjahr C, Paszkowski U (2013) Multiple control levels of root system remodeling in arbuscular mycorrhizal symbiosis. Front Plant Sci 4:204

Hennion NM, Durand C, Vriet, J et al (2019) Sugars en route to the roots. Transport, metabolism and storage within plant roots and towards microorganisms of the rhizosphere. Physiologia Plantarum 165(1):44–57

Hoang NV, Park C, Kamran M et al (2020a) Gene regulatory network guided investigations and engineering of storage root development in root crops. Front Plant Sci 11:762

Hoang NV, Choe G, Zheng Y et al (2020b) Identification of conserved gene-regulatory networks that integrate environmental sensing and growth in the root cambium. Curr Biol 30(15):2887–2900

Kays SJ (1985) The physiology of yield in the sweetpotato. In Bouwkamp J (ed) sweetpotato products: a natural resource for the tropics. CRC Press, pp 79–132

Keutgen N, Mukminah F, Roeb GW (2002) Sink strength and photosynthetic capacity influence tuber development in sweet potato. J Hortic Sci and Biotech 77(1):106–115

Kim SH, Hamada T, Otani M et al (2005) Cloning and characterization of sweetpotato MADS-box gene (IbAGL17) isolated from tuberous root. Plant Biotechnol J 22(3):217–220

Ku AT, Huang YS, Wang YS et al (2008) IbMADS1 (*I. batatas* MADS-box 1 gene) is involved in tuberous root initiation in sweet potato (*I. batatas*). Ann Bot 102(1):57–67

Li XQ, Zhang D (2003) Gene expression activity and pathway selection for sucrose metabolism in developing storage root of sweet potato. Plant Cell Physiol 44(6):630–636

McCormick FA (1916) Notes on the anatomy of the young tuber of Ipomoea batatas Lam. Bot Gaz 61(5):388–398

Miglietta F, Bindi M, Vaccari FP et al (2000) Crop ecosystem responses to climatic change: root and tuberous crops. In: Climate change and global crop productivity. CABI Publishing, Wallingford, UK, pp 189–212

Minchin PEH, Thorpe MR (1993) Sink strength: a misnomer, and best forgotten. Plant Cell Environ 16(9):1039–1040

Ngailo S, Shimelis H, Sibiya J et al (2019) Genotype-by-environment interaction of newly-developed sweet potato genotypes for storage root yield, yield-related traits and resistance to sweet potato virus disease. Heliyon 5(3):e01448

Noh SA, Lee HS, Huh EJ et al (2010) SRD1 is involved in the auxin-mediated initial thickening growth of storage root by enhancing proliferation of metaxylem and cambium cells in sweetpotato (*I. batatas*). J Exp Bot 61(5):1337–1349

Noh SA, Lee HS, Kim YS et al (2013) Down-regulation of the IbEXP1 gene enhanced storage root development in sweetpotato. J Exp Bot 64(1):129–142

Patrick JW (1993) Sink strength: whole plant considerations. Plant Cell Environ 16(9):1019–1020

Ravi V, Nascar SK, Makeshkumar T et al (2009) Molecular physiology of storage root formation and development in sweet potato (*Ipomoea batatas* (L.) Lam.). J Root Crops 35(1):1–27

Singh V, Sergeeva L, Ligterink W et al (2019) Gibberellin promotes sweetpotato root vascular lignification and reduces storage-root formation. Front Plant Sci 10:1320

Song W, Yan H, Ma M et al (2022) Comparative transcriptome profiling reveals the genes involved in storage root expansion in sweetpotato (*I. batatas* (L.) Lam.). Genes 3(7):1156

Stirling GR (2021) Modifying a productive sweet potato farming system in Australia to improve soil health and reduce losses from root-knot nematode. In: Sikora RA, Desaeger J (eds) Integrated nematode management: state-of-the-art and visions for the future. CABI, Wallingford, pp 368–373

Tanaka M (2016) Recent progress in molecular studies on storage root formation in sweetpotato (*Ipomoea batatas*). Jpn Agric Res Q 50(4):293–299

Tanaka M, Takahata Y, Nakatani M (2005). Analysis of genes developmentally regulated during storage root formation of sweet potato. J Plant Physiol 162(1):91–102

Thaler P, Pagès L (1998) Modelling the influence of assimilate availability on root growth and architecture. Plant Soil 201(2):307–320

Togari Y (1950) A study of tuberous root formation in sweet potato. Bull Nat Agr Expt Sta Tokyo 68:1–96

Tao X, Gu YH, Wang HY et al (2012) Digital gene expression analysis based on integrated de novo transcriptome assembly of sweet potato [*I. batatas* (L.) Lam.]. PLoS ONE 7(4):e36234

Villordon AQ, Ginzberg I, Firon N (2014) Root architecture and root and tuber crop productivity. Trends Plant Sci 19(7):419–425

Wang Z, Fang B, Chen X et al (2015) Temporal patterns of gene expression associated with tuberous root formation and development in sweetpotato (*I. batatas*). BMC Plant Biol 15(1):1–13

Wareing PF, Patrick J (1975) Source-sink relations and partition of assimilates in the plant. In: Cooper JP (ed) Photosynthesis and productivity in different environments. Cambridge Univ. Press, Cambridge, pp 481–499

Warren-Wilson J (1972) Control of crop processes. In: Rees AR, Cockshull KE, Hand DW et al (eds) Crop processes in controlled environment. Academic Press, London, pp 7–30

Wilson LA, Lowe SB (1973) The anatomy of the root system in West Indian sweet potato (*Ipomoea batatas* [L.] Lam.) cultivars. Ann Bot 37(3):633–643

Wolswinkel P (1985) Phloem unloading and turgor-sensitive transport: factors involved in sink control of assimilate partitioning. Physiol Plant 65(3):331–339

Yang J, An D, Zhang P (2011) Expression profiling of cassava storage roots reveals an active process of glycolysis/gluconeogenesis F. J Integr Plant Biol 53(3):193–211

Yang Y, Zhu J, Sun L et al (2023) Progress on physiological and molecular mechanisms of storage root formation and development in sweetpotato. Scientia Hortic 308:111588

You MK, Hur CG, Ahn YS et al (2003). Identification of genes possibly related to storage root induction in sweetpotato. FEBS letters, 536(1–3):101–105

Zierer W, Rüscher D, Sonnewald U et al (2021) Tuber and tuberous root development. Annu Rev Plant Biol 72(1):551–580

Zobel RW, Waisel YOAV (2010) A plant root system architectural taxonomy: a framework for root nomenclature. Plant Biosyst 144(2):507–512

Opportunities for Gene Editing of Sweetpotato

10

Debao Huang, Chase Livengood, G. Craig Yencho, and Wusheng Liu◉

Abstract

Sweetpotato plays significant roles in the food supply worldwide. Conventional sweetpotato breeding methods face challenges such as self- and cross-incompatibility and high heterogeneity. Gene editing is an effective and powerful tool for modifying agronomic traits, offering a novel approach to develop cultivars by targeting specific genes for precise modifications. The transformed CRISPR/Cas can be segregated out from the gene-edited end product of sexually propagated crops but not in sweetpotato as sweetpotato is highly heterogeneous and has to be propagated clonally. Thus, innovative sweetpotato breeding methods need to be further developed to improve breeding efficacy and decrease breeding cycle. In the present book chapter, we reviewed the methods used for sweetpotato breeding, the success of gene editing in sweetpotato, and the challenges and constraints and the future perspectives of sweetpotato gene editing.

Keywords

Gene editing · Sweetpotato

Sweetpotato (*Ipomoea batatas* (L.) Lam) is a globally important root, tuber, and banana (RTB) crop and cultivated in tropical and temperate zones of the world. As a hexaploidy crop with high heterogeneity, all the elite sweetpotato genotypes have to be propagated clonally, preventing further segregation of the genes and alleles combined in the elite genomes in the next generations. The asexual propagation of sweetpotatoes sometimes helps spread pathogens and pests, leading to dramatic yield loss. Sweetpotato breeders have to constantly develop new cultivars and lines with improved resistance to insects and pests that continuously evolve in the field. Sweetpotato breeders also need to put efforts to breed cultivars with improved yield, nutrition, and flavor due to market demands. Thus, sweetpotato breeding is of critical importance and innovative sweetpotato breeding methods need to be further developed in order to improve breeding efficacy and decrease breeding cycle and cost.

D. Huang · C. Livengood · G. C. Yencho · W. Liu (✉)
Department of Horticultural Science,
NC State University, Raleigh, NC 27607, USA
e-mail: wliu25@ncsu.edu

© The Author(s) 2025
G. C. Yencho et al. (eds.), *The Sweetpotato Genome*, Compendium of Plant Genomes,
https://doi.org/10.1007/978-3-031-65003-1_10

10.1 Methods Used for Sweetpotato Breeding

Three breeding methods are commonly employed in sweetpotato breeding programs, i.e., conventional breeding, molecular marker-assisted breeding, and transgenic breeding. As a hexaploid plant in the Convolvulaceae family, sweetpotato has the characteristics of self- and cross-incompatibility, high heterogeneity, and poor flowering and sterility in certain environments, which pose challenges for conventional sweetpotato breeding (Cervantes-Flores et al. 2011; Dhir et al. 1998; Yan et al. 2022). These challenges include the use of multiple parents for crossing, large (10,000–100,000) breeding populations for progeny evaluation, extended breeding cycles, genetic intricacies, unpredictable segregation, and labor-intensive procedures. Chromosomal linkage blocks and linkage-drag also prevent the generation of novel meiotic recombination of genes and alleles in conventional sweetpotato breeding. In the past few decades, both paired cross and polycross together with recurrent selection have been successfully implemented in conventional sweetpotato breeding for the development and release of elite cultivars with desirable traits. These traits include enhanced disease resistance, high yield, preferable taste, and improved nutrition such as high beta-carotene content, iron content, and dry matter, and low sweetness (Mwanga et al. 2016; Rolston et al. 1987).

Since conventional sweetpotato breeding is highly ineffective and time consuming, and conventional breeding for certain traits such as storage root yield and quality, resistance to root-knot nematode (RKN), sweetpotato virus disease (SPVD), and sweetpotato weevil has limited success (Collins et al. 2019; Ngailo et al. 2013; Oloka et al. 2021; Placide et al. 2015), molecular breeding has become a powerful and complementary means to conventional sweetpotato breeding for traits like these. Molecular breeding relies on the development of DNA-based molecular markers that are tightly linked to desirable traits to assist progeny

selection. Markers including Simple Sequence Repeats (SSRs), Restriction Fragment Length Polymorphisms (RFLPs), Amplified Fragment Length Polymorphisms (AFLPs), and more recently Single Nucleotide Polymorphisms (SNPs) have been used in marker-assisted selection (MAS) in sweetpotato (Chap. 4). MAS works well for simple traits and major QTLs. However, complex traits may be controlled by many small-effect genes and alleles. Molecular markers associated with each trait may be corresponding to a large genomic region containing tens or hundreds of genes rather than the genes responsible for the trait. Selection of markers is based on statistical significance of individual markers and sometimes arbitrary. To further improve selection efficiency and prediction of progeny performance, genomic selection can be used to estimate the effects of all the markers for all the traits of each individual plant to calculate genome-estimated breeding values (GEBVs) so that GEBVs can be used to predict which progeny is good for further testing and release. Efforts are underway in genomic selection in sweetpotato as variously described in Chaps. 4, 6, 7, and 12.

As the key genes underlying certain agronomic traits have been cloned and characterized in various species including sweetpotato, overexpression and/or RNAi-induced silencing of these key genes have been used in transgenic sweetpotato to improve various single gene traits. Thus, genetic engineering has emerged as a valuable new tool in sweetpotato breeding to improve salinity and drought tolerance, disease and pest resistance, herbicide resistance, and starch, carotenoid, and anthocyanin biosynthesis (Liu 2017). Both biolistic bombardment- and *Agrobacterium tumefaciens*-mediated transformation have been successfully used in the generation of transgenic sweetpotato lines with limited transformation efficiency (Liu 2017). Most of the transgenic sweetpotato lines were obtained from the transformation of leaves, petioles, stems, storage roots, and embryogenic calli even though transgenic sweetpotato plants were also generated from embryogenic suspension cell cultures. The advantage of using genetic engineering for sweetpotato improvement is the manipulation of one or a few genes at a time, permitting further finetuning of the expression of the gene(s) of interest. The difficulties in transgenic sweetpotato breeding include the genotype-dependency of sweetpotato transformation and the public acceptance of the transgenic end products. Most elite sweetpotato breeding cultivars and lines are not transformable and the transformable genotypes are not elite. The permanent integration of the transgenes in the end products makes the sweetpotatoes GMO, which is a marketing hurdle.

In addition to the above-mentioned breeding methods, somatic hybridization (Guo et al. 2006; Yang et al. 2009; Zhang et al. 2001) and mutation breeding (Mansour et al., 2018; Moussa and Gomaa 2017; Luan et al. 2007; Shin et al. 2011; Wang et al. 2007; Yan et al. 2022) have been successfully used in sweetpotato breeding. Somatic hybrids were successfully obtained between sweetpotato and its wild relatives *I. cairica* (Guo et al. 2006), *I. lacunose* (Zhang et al. 2001), or *I. triloba* (Jia et al. 2022) even though many of these somatic hybrids contained significantly reduced chromosome numbers than the sum of the chromosome numbers in both parents. Using gamma irradiation, improved sweetpotato varieties have been developed from axillary buds for high yield and starch content (Shin et al. 2011), shoot apices for changed root flesh color and increased root yield (Wang et al. 2007), stems for modified yield and quality traits (Moussa and Gomaa, 2017), and calli for induced morphological changes (Lee et al. 2002). The use of gamma irradiation for sweetpotato breeding may suffer from the formation of chimera mutations, which can be avoided in the ethylmethanesulphonate (EMS)-mediated sweetpotato breeding that utilized sweetpotato leaf explant-derived calli to breed cultivars with enhanced salt tolerance (Luan et al. 2007). All of these mutation breeding methods were used in combination with tissue culture and may suffer from a low mutation rate.

10.2 Gene Editing Biotechnologies that Could Be Applied in Sweetpotato

Gene editing tools include meganucleases (Cohen-Tannoudji et al. 1998), zinc-finger nuclease (ZFNs) (Bibikova et al. 2002), transcription activator-like effector nucleases (TALENs) (Christian et al. 2010), and clustered regularly interspaced short palindromic repeat-associated nuclease (CRISPR) (Cong et al. 2013; Mali et al. 2013). These nucleases cut both DNA strands at the target sites to create double-strand breaks (DSBs). The DSBs are then repaired by the DNA repair mechanisms in plant cells. The dominant repair mechanism in plant cells is non-homologous end joining (NHEJ), which is prone to introducing random mutations. DSBs can also be repaired to a lesser extent by homology-directed repair (HDR) if a donor DNA template is provided for homologous recombination. Screening of the mutated genes can lead to the identification of the gene-edited plants.

Meganucleases have been employed as a gene editing tool since 1985. They work as a homodimer to bind their specific recognition sites ranging from 12 to 40 base pairs in length (Fig. 10.1a; Jurica et al. 1998). The rare presence of their recognition sites in plant genomes and the fusion of the DNA-binding domain with their catalytic domain limit their application in plants. In the 1990s, ZFNs emerged as a promising gene editing tool. ZFNs work as heterodimers with each ZFN containing a specific DNA-binding domain and a non-specific cleavage domain from the Fok I endonuclease (Fig. 10.1b; Pabo et al. 2001). Each DNA-binding domain contains 6–8 zinc fingers with each zinc finger recognizing and binding to 3-bp-long nucleotides. Design of each DNA-binding domain requires screening against expression libraries to confirm their binding specificity. Synthesis of each DNA-binding domain is tedious and costly. The prohibitive cost and intricate synthesis process have hampered the widespread adoption of ZFNs in plant applications. Subsequently, TALENs were developed, featuring another type of DNA-binding domain and a non-specific Fok I domain positioned at the carboxylic terminal (Fig. 10.1c; Christian et al. 2010). The DNA-binding domain of TALENS contains ~20 TALE repeats with each repeat identical in amino acid sequence except the Repeat Variable Diresidue (RVD) on positions 12 and 13 of each repeat that binds to a single nucleotide (Christian et al. 2010). Compared to ZFNs, TALENs offer greater user-friendliness, although their drawback lies in the necessity of constructing each repeat, which is costly.

More recently, CRISPR/Cas technologies have risen in prominence due to their efficiency and ease of use. Each CRISPR/Cas consists of two primary components, i.e., an endonuclease Cas protein and a single guide RNA (gRNA), which binds to the single-strand DNA through 17–20 nt at the 5'-end of each gRNA (Fig. 10.1d; Tsai et al. 2015). With the guidance of a gRNA, the Cas protein specifically binds to and cleaves the target DNA, triggering DNA repair. There are various types of Cas genes such as *Cas9* (Chen et al. 2019; Zhu et al. 2020), *Cas12a* (Wada et al. 2022), and *Cas13a* (Abudayyeh et al. 2016, 2019; East-Seletsky et al. 2016; Konermann et al. 2018). Cas9/gRNA stands as a highly efficient tool that has been extensively used for precise modification of target genes in plants. It has been employed for gene disruption, gene insertion, gene replacement, base editing, and regulation of gene expression in crops. It has played a pivotal role in modifying agronomic traits, encompassing improvements in nutritional content, yield enhancement, and the development of stress and disease-resistant crops (Zhu et al. 2020; Gao 2021). It expedites crop breeding by facilitating the incorporation of desired traits while minimizing undesirable ones.

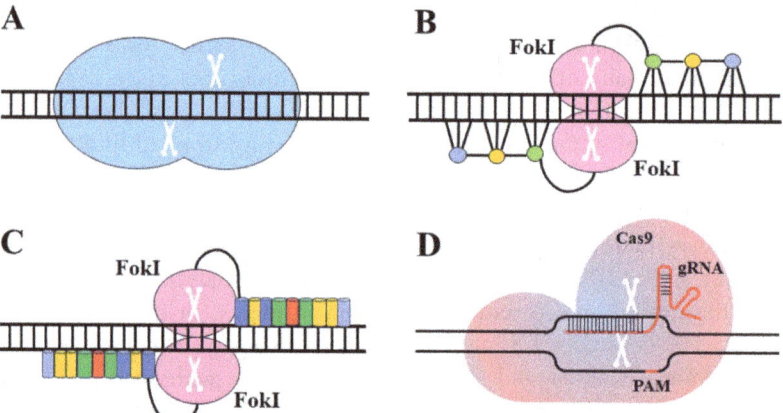

Fig. 10.1 Schematic representations of gene editing tools. **a** Meganuclease can bind 12–40-bp-long DNA sequences and precisely cleave both strands at its recognition site, resulting in sticky DSBs. **b** Zinc-finger nucleases (ZFNs) function as dimers, with each monomer comprising a DNA-binding domain and a nuclease domain. The DNA-binding domain comprises 3–6 zinc finger repeats, forming an array that identifies 9–18 nucleotides. The nuclease domain contains the type II restriction endonuclease Fok I. **c** Transcription activator-like nucleases (TALENs) operate as dimeric enzymes akin to ZFNs. Each subunit contains a DNA-binding domain—a highly conserved 33–34-amino acid-long sequence tailored for each nucleotide, and a Fok I nuclease domain. **d** In the CRISPR/Cas9 system, the Cas9 endonuclease is directed by the gRNA to achieve precise target cleavage. A 20-nucleotide-long recognition site precedes the protospacer adjacent motif (PAM) for this process

10.3 The Success of Gene Editing in Sweetpotato

To date, the CRISPR/Cas technology has been used in sweetpotato for gene editing while the other gene editing tools have never been reported for attempts in sweetpotato. CRISPR/Cas9 was used to knockout the granule-bound starch synthase I (*GBSSI*) gene controlling amylose biosynthesis and starch branching enzyme II (*SBEII*) gene responsible for amylopectin biosynthesis, and a mutation efficiency of 62–92% was achieved for multi-allelic mutations (Wang et al. 2019). Most of the detected mutations were point mutations and small insertions/deletions (indels), some of which caused amino acid changes or stop codons. The *gbssI* knockout showed reduced amylose content while the *sbeII* knockout had increased amylose content and decreased amylopectin content. Neither individual knockout caused significant changes in the total starch content. This pioneer example demonstrated the effectiveness of the use of CRISPR/Cas9 for gene editing and trait improvement in sweetpotato.

Moreover, Cao et al. (2022) successfully obtained the *PDS* mutant through the cut-dip-budding delivery system in several sweetpotato genotypes. The cut-dip-budding method was developed based on the "root suckering" ability of sweetpotatoes, which grows shoots from adventitious shoot primordia on roots. In this process, the shoot explants were cut off at the shoot–root junction and treated with *A. rhizogenes* containing *Cas9/gRNA* at the cut sites to induce transformed roots, followed by bud generation. Mei et al. (2024) also reported that they successfully obtained the *PDS* mutant through the gene editing with the injection delivery method in sweetpotatoes. The axillary buds were removed from sweetpotato shoots, and *A. tumefaciens* containing *Cas9/gRNA* was injected into the cut sites for the regeneration of gene-edited shoots. The use of gene editing for sweetpotato trait improvement provides precision in the modification of target genes. It also decreases the sweetpotato breeding cycle from 8 to 12 years for conventional breeding to just one generation, significantly accelerating sweetpotato breeding and trait improvement.

10.4 Challenges and Constraints in Sweetpotato Gene Editing

To conduct gene editing in sweetpotato, Cas/gRNA needs to be delivered into a sweetpotato genome, which is typically achieved through biolistic bombardment- or *Agrobacterium*-mediated stable sweetpotato transformation. However, sweetpotato transformation is highly genotype dependent. Most, if not all, of the elite sweetpotato cultivars and lines are not transformable, preventing the direct use of gene editing in these elite sweetpotato cultivars and lines for further trait improvement.

The transformation and permanent integration of Cas/gRNA into a sweetpotato genome makes the end product GMO. Unlike diploid crops such as tomato in which sexual propagation and genetic segregation can be used to segregate out the transgenes, segregating out the transgenes via sexual propagation cannot be applied in sweetpotato. As a highly heterogeneous hexaploid crop, elite sweetpotato breeding cultivars and lines can only be propagated asexually once all the favorable genes and alleles have been combined into an individual plant through crossing. This constraint prevents the removal of the transgenes from the end product via segregation following sweetpotato transformation and gene editing, making the end product GMO. GMO gene-edited sweetpotato suffers from public acceptance and marketing hurdles in some countries such as England and the E.U.

Challenges and constraints in sweetpotato gene editing also come from the incomplete whole genome sequences of limited sweetpotato cultivars and their hexaploid nature, making gRNA target design challenging for genes with unknown sequences or with many homologous sequences in a sweetpotato genome of interest. The first successful attempt to conduct whole genome sequencing in sweetpotato was reported in Yang et al. (2017) which published a half haplotype-resolved genome in a newly bred carotenoid-rich cultivar Taizhong6. Then, Yoon et al. (2022) published a haploid-resolved and chromosome-scale assembly of the whole genome sequence of sweetpotato cv. Xushu18, which identified 175,633 genes and suggested that cv. Xushu18 is an auto-hexaploid with an AAAAAB genome. In addition, investments by the Bill & Melinda Gates Foundation in the GT4SP (http://sweetpotato.uga.edu/gt4sp_download.shtml) and SweetGAINS (https://cipotato.org/cip_projects/sweetgains-africa/) projects have developed genomics tools for sweetpotato improvement and have significantly improved the sweetpotato genomic resources, respectively. In August 2022, the SweetGAINS project released the high-quality v1 genome assembly of cv. Beauregard with annotation that is now available for searches with the BLAST search tool (http://sweetpotato.uga.edu/blast.shtml). In addition, the complete chloroplast genome sequences have been reported for 16 sweetpotato cultivars (Zhou et al. 2018) and another 107 sweetpotato cultivars including cv. Xushu18 (Xiao et al. 2021; Yan et al. 2015; Yoon et al. 2022). The whole mitochondrial genomes have also been published for cv. Xushu18 (Yoon et al. 2022) and cv. JinShan 57 (Yang et al. 2022). The availability of these genomic sequences dramatically helps with gRNA design for gene editing in sweetpotato. However, there may exist various SNPs between the published cultivar genomes and cultivar genomes of interest, which may affect the accuracy of the designed gRNA target sites and decrease the editing efficiency with potential off-target effects and unintended consequences in cultivar genomes of interest. To improve gRNA design, PCR amplification followed by Sanger sequencing could be used to amplify a gene of interest from a cultivar genome of interest even though it may be challenging to amplify all the six alleles of a gene in a cultivar genome.

10.5 Future Perspectives of Gene Editing in Sweetpotato

To overcome the above-mentioned challenges and constraints of sweetpotato gene editing, innovative approaches for genotype-independent transgene-free gene editing in crops including sweetpotato need to be developed. A set of key growth and developmental regulatory genes have been recently found to make recalcitrant crop genotypes transformable, permitting genotype-independent transformation in certain groups of crops (for reviews, see Gordon-Kamm et al. 2019; Maren et al. 2022; Nagle et al. 2018; Nalapalli et al. 2021). The most well-known growth and developmental regulatory genes include the *WUSCHEL* (*WUS*), *BABY BOOM* (*BBM*), *ISOPENTENYL TRANSFERASE* (*IPT*), *GROWTH-REGULATING FACTOR* (*GRF4*), and *GRF-INTERACTING FACTOR1* (*GIF1*) genes. A pioneering example came from the use of a low expression of the maize *WUS2* gene and a constitutive expression of the maize *BBM* gene for genotype-independent crop transformation in four difficult-to-transform maize inbred lines as well as thirty-three commercial maize inbred lines (Lowe et al. 2016). Another pioneering example was published in Debernardi et al. (2020) for the use of the wheat *GRF4-GIF1* chimeric gene for genotype-independent transformation in wheat, rice, citrus, and triticale. Moreover, Maher et al. (2020) reported the use of a low expression of the maize *WUS2* gene and a high expression of the *Agrobacterium IPT* gene for Cas9-mediated gene editing and edited shoot regeneration from the mature plants of tobacco, potato, and grape, which avoided the use of plant tissue culture. All these examples offer a great promise for using growth and developmental regulatory genes to conduct genotype-independent transformation in sweetpotato for gene editing.

To date, several strategies have been developed for transgene-free gene editing in crops. These include transgene removal via genetic segregation, transient expression of the *Cas/gRNA* DNA without permanent integration of *Cas/gRNA* into crop genomes, and DNA-free (or protein) delivery of the pre-assembled Cas/gRNA ribonucleoproteins (RNPs) or pre-transcribed Cas/gRNA RNA. As previously mentioned, elite sweetpotato cultivars must be propagated asexually to maintain favorable alleles, thus preventing the removal of transgenes via genetic segregation.

Cas/gRNA, delivered by either particle bombardment- or *Agrobacterium*-mediated transformation, can be transiently expressed in the nucleus without stable integration into the host genome. Thus, transgene-free gene-edited plants can be regenerated and selected in the absence of a selection agent; these regenerated plants will contain the targeted edited gene(s) but will not contain the transgenes. Identifying the transgene-free gene-edited plants requires a highly efficient plant regeneration protocol and a large number of regenerated plants after transformation. Simultaneous editing of a gene of interest (GOI) and the acetolactate synthase gene (*ALS*) gene in tomato, tobacco, potato, and citrus provided a selection marker for edited cells as the mutated *als* gene conferred resistance to sulfonylurea herbicides (Huang et al. 2023). While this is an effective method for obtaining transgene-free gene-edited plants, it still relies upon crop transformation for DNA delivery, which remains an obstacle for sweetpotato.

Moreover, Cas/gRNA RNPs can be delivered into a crop genome in a DNA-free manner, bypassing crop transformation of the Cas9/gRNA DNA. Protein delivery of the Cas/gRNA RNP has been used to obtain transgene-free gene-edited plants in a variety of crop species using biolistic bombardment (Liang et al. 2018, 2019; Poddar et al. 2023; Svitashev et al. 2016), polyethyleneglycol (PEG)-mediated transfection (Andersson et al. 2018; Banakar et al. 2022; Brandt et al. 2020; Choi et al. 2021; Fan et al. 2020; Jiang et al. 2021; Kim et al. 2020a, b; Klimek-Chodacka et al. 2021; Lin et al. 2022; Malnoy et al. 2016; Murovec et al. 2018; Nicolia et al. 2021; Najafi et al. 2023; Park et al. 2019; Pavese et al. 2022; Sidorov et al. 2022;

Subburaj et al. 2016; Yu et al. 2021; Woo et al. 2015; Wu et al. 2020), and lipofection-mediated transfection (Liu et al. 2020). Most methods utilize protoplasts, which have been stripped of their cell walls for an easy delivery of RNPs across the cell membranes. However, most crop species, including sweetpotato, do not have a well-established plant regeneration system from protoplasts. Although protoplast regeneration has been previously reported in sweetpotato, efficiency appears to be genotype dependent (Sihachakr and Ducreux 1987). Protoplast-based systems also tend to be more technically challenging than other methods, requiring specialized equipment and labor.

Delivery of gene editing RNA transcripts into plant cells has also been used to achieve transgene-free gene editing in plants. Biolistic bombardment and virus-mediated delivery methods have been used to deliver Cas9/gRNA transcripts into plant cells for transient editing (Ma et al. 2020; Zhang et al. 2016). A recently published grafting-based system delivered Cas9/gRNA transcripts from transgenic rootstocks to wild-type scions using tRNA-like sequence (TLS) motifs (Yang et al. 2023). The addition of TLS motifs to Cas9 and gRNA transcripts allows them to traverse graft unions, resulting in transgene-free gene editing of the wild-type scion tissues. Like the DNA-based transient transformation systems outlined earlier, cells that have been edited using these methods must also be regenerated in a selection-free environment, making it difficult to identify edited plants.

Taken together, gene editing is a promising and powerful bioengineering method and can be used together with other breeding techniques to improve sweetpotato traits as well as conduct gene functional analysis. Enabling tools such as genotype-independent transgene-free gene editing methods need to be developed to revolutionize sweetpotato breeding.

Acknowledgements This work was supported by the Hatch projects 02779 and 02685 from the US Department of Agriculture National Institute of Food and Agriculture.

Conflicts of Interest The authors declare no conflict of interest.

References

Abudayyeh OO, Gootenberg JS, Konermann S, Joung J, Slaymaker IM, Cox DB, Shmakov S, Makarova KS, Semenova E, Minakhin L (2016) C2c2 is a single-component programmable RNA-guided RNA-targeting CRISPR effector. Science 353:aaf5573

Abudayyeh OO, Gootenberg JS, Kellner MJ, Zhang F (2019) Nucleic acid detection of plant genes using CRISPR-Cas13. CRISPR J 2:165–171

Andersson M, Turesson H, Olsson N, Fält AS, Ohlsson P, Gonzalez MN, Samuelsson M, Hofvander P (2018) Genome editing in potato via CRISPR-Cas9 ribonucleoprotein delivery. Physiol Plant 164:378–384

Banakar R, Rai KM, Zhang F (2022) CRISPR DNA-and RNP-mediated genome editing genome editing via *Nicotiana benthamiana* protoplast transformation protoplast transformation and regeneration protoplast regeneration. Protoplast technology: methods and protocols. Springer, New York, pp 65–82

Bibikova M, Golic M, Golic KG, Carroll D (2002) Targeted chromosomal cleavage and mutagenesis in *Drosophila* using zincfinger nucleases. Genetics 161:1169–1175

Brandt KM, Gunn H, Moretti N, Zemetra RS (2020) A streamlined protocol for wheat (*Triticum aestivum*) protoplast isolation and transformation with CRISPR-Cas ribonucleoprotein complexes. Front Plant Sci 11:769

Cao X, Xie H, Song M, Lu J, Ma P, Huang B, Wang M, Tian Y, Chen F, Peng J, Lang Z, Li G, Zhu JK (2022) Cut-dip-budding delivery system enables genetic modifications in plants without tissue culture. Innovation 4:100345

Cervantes-Flores JC, Sosinski B, Pecota KV, Mwanga ROM, Catignani GL, Truong VD, Watkins RH, Ulmer MR, Yencho GC (2011) Identification of quantitative trait loci for dry-matter, starch, and β-carotene content in sweetpotato. Mol Breed 28:201–216

Chen K, Wang Y, Zhang R, Zhang H, Gao C (2019) CRISPR/Cas genome editing and precision plant breeding in agriculture. Annu Rev Plant Biol 70:667–697

Choi SH, Lee MH, Jin DM, Ju SJ, Ahn WS, Jie EY, Lee JM, Lee J, Kim CY, Kim SW (2021) TSA promotes CRISPR/Cas9 editing efficiency and expression of cell division-related genes from plant protoplasts. Int J Mol Sci 22:7817

Christian M, Cermak T, Doyle EL, Schmidt C, Zhang F, Hummel A, Bogdanove AJ, Voytas DF (2010) Targeting DNA double-strand breaks with TAL effector nucleases. Genetics 186:757–776

Cohen-Tannoudji M, Robine S, Choulika A, Pinto D, El Marjou F, Babinet C, Louvard D, Jaisser F (1998) I-SceI-induced gene replacement at a natural locus in embryonic stem cells. Mol Cell Biol 18:1444–1448

Collins WW, Jones A, Mullen MA, Talekar NS, Martin FW (2019) Breeding sweet potato for insect resistance: a global overview. Sweet Potato Pest Manag 12:379–397

Cong L, Ran FA, Cox D, Lin S, Barretto R, Habib N, Hsu PD, Wu X, Jiang W, Marraffini LA, Zhang F (2013) Multiplex genome engineering using CRISPR/Cas systems. Science 339:819–823

Debernardi JM, Tricoli DM, Ercoli MF, Hayta S, Ronald P, Palatnik JF, Dubcovsky J (2020) A GRF–GIF chimeric protein improves the regeneration efficiency of transgenic plants. Nat Biotechnol 38:1274–1279

Dhir SK, Oglesby J, Bhagsari AS (1998) Plant regeneration via somatic embryogenesis, and transient gene expression in sweetpotato protoplasts. Plant Cell Rep 17:665–669

East-Seletsky A, O'Connell MR, Knight SC, Burstein D, Cate JH, Tjian R, Doudna JA (2016) Two distinct RNase activities of CRISPR-C2c2 enable guide-RNA processing and RNA detection. Nature 538:270–273

Fan Y, Xin S, Dai X, Yang X, Huang H, Hua Y (2020) Efficient genome editing of rubber tree (*Hevea brasiliensis*) protoplasts using CRISPR/Cas9 ribonucleoproteins. Ind Crops Prod 146:112146

Gao C (2021) Genome engineering for crop improvement and future agriculture. Cell 184:1621–1635

Gordon-Kamm B, Sardesai N, Arling M, Lowe K, Hoerster G, Betts S, Jones T (2019) Using morphogenic genes to improve recovery and regeneration of transgenic plants. Plants 8:38

Guo JM, Liu QC, Zhai H, Wang YP (2006) Regeneration of plants from *Ipomoea cairica* L protoplasts and production of somatic hybrids between *I. cairica* L. and sweetpotato, *I. batatas* (L.) Lam. Plant Cell Tissue Organ Cult 87:321–327

Huang X, Jia H, Xu J, Wang Y, Wen J, Wang N (2023) Transgene-free genome editing of vegetatively propagated and perennial plant species in the T_0 generation via a co-editing strategy. Nat Plants 9:1591–1597

Jia L, Yang Y, Zhai H, He S, Xin G, Zhao N, Zhang H, Gao S, Liu Q (2022) Production and characterization of a novel interspecific somatic hybrid combining drought tolerance and high quality of sweet potato and *Ipomoea triloba* L. Plant Cell Rep 41:2159–2171

Jiang W, Bush J, Sheen J (2021) A versatile and efficient plant protoplast platform for genome editing by Cas9 RNPs. Front Genome Editing 3:719190

Jurica MS, Monnat RJ Jr, Stoddard BL (1998) DNA recognition and cleavage by the LAGLIDADG homing endonuclease I-CreI. Mol Cell 2:469–476

Kim H, Choi J, Won KH (2020a) A stable DNA-free screening system for CRISPR/RNPs-mediated gene editing in hot and sweet cultivars of *Capsicum annuum*. BMC Plant Biol 20:1–12

Kim HS, Wang WB, Kang L, Kim SE, Lee CJ, Park SC, Park WS, Ahn MJ, Kwak SS (2020b) Metabolic engineering of low-molecular-weight antioxidants in sweetpotato. Plant Biotechnol Rep 14:193–205

Klimek-Chodacka M, Gieniec M, Baranski R (2021) Multiplex site-directed gene editing using polyethylene glycol-mediated delivery of CRISPR gRNA: Cas9 ribonucleoprotein (RNP) complexes to carrot protoplasts. Int J Mol Sci 22:10740

Konermann S, Lotfy P, Brideau NJ, Oki J, Shokhirev MN, Hsu PD (2018) Transcriptome engineering with RNA-targeting type VI-D CRISPR effectors. Cell 173:665–676

Lee YI, Lee IS, Lim YP (2002) Variations in sweetpotato regenerates from gamma-ray irradiated embryogenic callus. J Plant Biotechnol 4:163–170

Liang Z, Chen K, Zhang Y, Liu J, Yin K, Qiu JL, Gao C (2018) Genome editing of bread wheat using biolistic delivery of CRISPR/Cas9 in vitro transcripts or ribonucleoproteins. Nat Protoc 13:413–430

Liang Z, Chen K, Gao C (2019) Biolistic delivery of CRISPR/Cas9 with ribonucleoprotein complex in wheat. In: Plant genome editing with CRISPR systems: methods and protocols, pp 327–335

Lin CS, Hsu CT, Yuan YH, Zheng PX, Wu FH, Cheng QW, Wu YL, Wu TL, Lin S, Yue JJ, Cheng YH (2022) DNA-free CRISPR-Cas9 gene editing of wild tetraploid tomato Solanum peruvianum using protoplast regeneration. Plant Physiol 188:1917–1930

Liu Q (2017) Improvement for agronomically important traits by gene engineering in sweetpotato. Breeding Sci 67:15–26

Liu W, Rudis MR, Cheplick MH, Millwood RJ, Yang J, Ondzighi-Assoume CA, Montgomery GA, Burris KP, Mazarei M, Chesnut JD, Stewart CN (2020) Lipofection-mediated genome editing using DNA-free delivery of the Cas9/gRNA ribonucleoprotein into plant cells. Plant Cell Rep 39:245–257

Lowe K, Wu E, Wang N, Hoerster G, Hastings C, Cho MJ, Scelonge C, Lenderts B, Chamberlin M, Cushatt J, Wang L (2016) Morphogenic regulators Baby boom and Wuschel improve monocot transformation. Plant Cell 28:1998–2015

Luan YS, Zhang J, Gao XR, An LJ (2007) Mutation induced by ethylmethanesulphonate (EMS), in vitro screening for salt tolerance and plant regeneration of sweet potato (Ipomoea batatas L). Plant Cell Tissue Organ Cult 88:77–81

Ma X, Zhang X, Liu H, Li Z (2020) Highly efficient DNA-free plant genome editing using virally delivered CRISPR-Cas9. Nat Plants 6:773–779

Maher MF, Nasti RA, Vollbrecht M, Starker CG, Clark MD, Voytas DF (2020) Plant gene editing through de novo induction of meristems. Nat Biotechnol 38:84–89

Mali P, Yang L, Esvelt KM, Aach J, Guell M, DiCarlo JE, Norville JE, Church GM (2013) RNA-guided human genome engineering via Cas9. Science 339:823–826

Malnoy M, Viola R, Jung MH, Koo OJ, Kim S, Kim JS, Velasco R, Kanchiswamy NC (2016) DNA-free genetically edited grapevine and apple protoplast using CRISPR/Cas9 ribonucleoproteins. Front Plant Sci 7:1904

Mansour MK, Abido A, Yousry M, Moussa S (2018) Selection of new distinct sweet potato clones using chemical mutagen agents and gamma-ray radiation. J Adv Agric Res 23:168–193

Maren NA, Duan H, Da K, Yencho GC, Ranney TG, Liu W (2022) Genotype-independent plant transformation. Hortic Res 9:uhac047

Mei G, Chen A, Wang Y, Li S, Wu M, Hu Y, Liu X, Hou X (2024) A simple and efficient in planta transformation method based on the active regeneration capacity of plants. Plant Com 5:100822

Moussa SA, Gomaa SE (2017) Mutation breeding and assessment of clones induced through gamma radiation of the sweet potato cultivar (Abees). Egypt J Plant Breed 21:312–338

Murovec J, Guček K, Bohanec B, Avbelj M, Jerala R (2018) DNA-free genome editing of Brassica oleracea and B. rapa protoplasts using CRISPR-Cas9 ribonucleoprotein complexes. Front Plant Sci 9:1594

Mwanga RO, Kyalo G, Ssemakula GN, Niringiye C, Yada B, Otema MA, Namakula J, Alajo A, Kigozi B, Makumbi RN, Ball A, Grüneberg WJ, Low JW, Yencho GC (2016) 'NASPOT 12 O' and 'NASPOT 13 O' sweetpotato. HortScience 51:291–295

Nagle M, Déjardin A, Pilate G, Strauss SH (2018) Opportunities for innovation in genetic transformation of forest trees. Front Plant Sci 9:1443

Najafi S, Bertini E, D'Incà E, Fasoli M, Zenoni S (2023) DNA-free genome editing in grapevine using CRISPR/Cas9 ribonucleoprotein complexes followed by protoplast regeneration. Hortic Res 10:uhac240

Nalapalli S, Tunc-Ozdemir M, Sun Y, Elumalai S, Que Q (2021) Morphogenic regulators and their application in improving plant transformation. In: Bandyopadhyay A, Thilmony R (eds) Rice genome engineering and gene editing. Methods in molecular biology, pp 37–61

Ngailo S, Shimelis H, Sibiya J, Mtunda K (2013) Sweet potato breeding for resistance to sweet potato virus disease and improved yield: progress and challenges. Afr J Agric Res 8:3202–3215

Nicolia A, Andersson M, Hofvander P, Festa G, Cardi T (2021) Tomato protoplasts as cell target for ribonucleoprotein (RNP)-mediated multiplexed genome editing. Plant Cell Tissue Organ Cult 144:463–467

Oloka BM, da Silva Pereira G, Amankwaah VA, Mollinari M, Pecota KV, Yada B, Olukolu BA, Zeng ZB, Yencho CG (2021) Discovery of a major QTL for root-knot nematode (Meloidogyne incognita) resistance in cultivated sweetpotato (Ipomoea batatas). Theor Appl Genet 134:1945–1955

Pabo CO, Peisach E, Grant RA (2001) Design and selection of novel Cys2His2 zinc finger proteins. Annu Rev Biochem 70:313–340

Park, J, Choi, S, Park, S, Yoon, J, Park, AY, and Choe, S (2019) DNA-free genome editing via ribonucleoprotein (RNP) delivery of CRISPR/Cas in lettuce. In: Plant genome editing with CRISPR systems: methods and protocols, pp 337–354

Pavese V, Moglia A, Abbà S, Milani AM, Torello Marinoni D, Corredoira E, Martínez MT, Botta R (2022) First report on genome editing via ribonucleoprotein (RNP) in *Castanea sativa*. Mill Int J Mol Sci 23:5762

Placide R, Shimelis H, Laing M, Gahakwa D (2015) Application of principal component analysis to yield and yield related traits to identify sweet potato breeding parents. Trop Agric 92:1–15

Poddar S, Tanaka J, Running KL, Kariyawasam GK, Faris JD, Friesen TL, Cho MJ, Cate JH, Staskawicz B (2023) Optimization of highly efficient exogenous-DNA-free Cas9-ribonucleoprotein mediated gene editing in disease susceptibility loci in wheat (*Triticum aestivum* L). Front Plant Sci 13:1084700

Rolston LH, Clark CA, Cannon JM, Randle WM, Riley EG, Wilson PW, Robbins ML (1987) 'Beauregard' sweet potato. HortScience 22:1338–1339

Shin JM, Kim B, Seo S, Jeon SB, Kim J, Jun B, Kang S, Lee JS, Chung M, Kim SB (2011) Mutation breeding of sweet potato by gamma-ray radiation. Afr J Agric Res 6:1447–1454

Sidorov V, Wang D, Nagy ED, Armstrong C, Beach S, Zhang Y, Groat J, Yang S, Yang P, Gilbertson L (2022) Heritable DNA-free genome editing of canola (*Brassica napus* L) using PEG-mediated transfection of isolated protoplasts. In Vitro Cell Dev Biol 58:447–456

Sihachakr D, Ducreux G (1987) Plant regeneration from protoplast culture of sweet potato (*Ipomoea batatas* Lam). Plant Cell Rep 6:326–328

Subburaj S, Chung SJ, Lee C, Ryu SM, Kim DH, Kim JS, Bae S, Lee GJ (2016) Site-directed mutagenesis in *Petunia× hybrida* protoplast system using direct delivery of purified recombinant Cas9 ribonucleoproteins. Plant Cell Rep 35:1535–1544

Svitashev S, Schwartz C, Lenderts B, Young JK, Cigan AM (2016) Genome editing in maize directed by CRISPR-Cas9 ribonucleoprotein complexes. Nat Commun 1:13274

Tsai SQ, Zheng Z, Nguyen NT, Liebers M, Topkar VV, Thapar V, Wyvekens N, Khayter C, Iafrate AJ, Le LP, Aryee MJ, Joung JK (2015) GUIDE-seq enables genome-wide profiling of off-target cleavage by CRISPR-Cas nucleases. Nat Biotechnol 33:187–197

Wada N, Osakabe K, Osakabe Y (2022) Expanding the plant genome editing toolbox with recently developed CRISPR-Cas systems. Plant Physiol 188:1825–1837

Wang Y, Wang F, Zhai H, Liu Q (2007) Production of a useful mutant by chronic irradiation in sweetpotato. Sci Hortic 111:173–178

Wang H, Wu Y, Zhang Y, Yang J, Fan W, Zhang H, Zhao S, Yuan L, Zhang P (2019) CRISPR/Cas9-based mutagenesis of starch biosynthetic genes in sweet potato (*Ipomoea Batatas*) for the improvement of starch quality. Int J Mol Sci 20:4702

Woo JW, Kim J, Kwon SI, Corvalán C, Cho SW, Kim H, Kim S, Kim S, Choe S, Kim J (2015) DNA-free genome editing in plants with preassembled CRISPR-Cas9 ribonucleoproteins. Nat Biotechnol 33:1162–1164

Wu S, Zhu H, Liu J, Yang Q, Shao X, Bi F, Hu C, Huo H, Chen K, Yi G (2020) Establishment of a PEG-mediated protoplast transformation system based on DNA and CRISPR/Cas9 ribonucleoprotein complexes for banana. BMC Plant Biol 20:1–10

Xiao S, Xu P, Deng Y, Dai X, Zhao L, Heider B, Zhang A, Zhou Z, Cao Q (2021) Comparative analysis of chloroplast genomes of cultivars and wild species of sweetpotato (*Ipomoea batatas* [L] Lam). BMC Genom 22:1–12

Yan L, Lai X, Li X, Wei C, Tan X, Zhang Y (2015) Analyses of the complete genome and gene expression of chloroplast of sweet potato (*Ipomoea batata*). PLoS ONE 10:e0124083

Yan M, Nie H, Wang Y, Wang X, Jarret R, Zhao J, Wang H, Yang J (2022) Exploring and exploiting genetics and genomics for sweetpotato improvement: status and perspectives. Plant Commun 3:100332

Yang Y, Guan S, Zhai H, He S, Liu Q (2009) Development and evaluation of a storage root-bearing sweetpotato somatic hybrid between *Ipomoea batatas* (L) Lam and *I triloba* L. Plant Cell Tissue Organ Cult 99:83–89

Yang J, Moeinzadeh MH, Kuhl H, Helmuth J, Xiao P, Haas S, Liu G, Zheng J, Sun Z, Fan W, Deng G (2017) Haplotype-resolved sweet potato genome traces back its hexaploidization history. Nat Plants 3:696–703

Yang Z, Ni Y, Lin Z, Yang L, Chen G, Nijiati N, Hu Y, Chen X (2022) De novo assembly of the complete mitochondrial genome of sweet potato (*Ipomoea batatas* [L] Lam) revealed the existence of homologous conformations generated by the repeat-mediated recombination. BMC Plant Biol 22:285

Yang L, Machin F, Wang S, Saplaoura E, Kragler F (2023) Heritable transgene-free genome editing in plants by grafting of wild-type shoots to transgenic donor rootstocks. Nat Biotechnol 41:958–967

Yoon UH, Cao Q, Shirasawa K, Zhai H, Lee TH, Tanaka M, Hirakawa H, Hahn JH, Wang X, Kim HS, Tabuchi H et al (2022) Haploid-resolved and chromosome-scale genome assembly in hexa-autoploid sweetpotato (*Ipomoea batatas* (L) Lam). bioRxiv, 2022–12

Yu J, Tu L, Subburaj S, Bae S, Lee GJ (2021) Simultaneous targeting of duplicated genes in *Petunia* protoplasts for flower color modification via CRISPR-Cas9 ribonucleoproteins. Plant Cell Rep 40:1037–1045

Zhang BY, Liu QC, Zhai H, Zhou HY (2001) Production of fertile interspecific somatic hybrid plants between sweetpotato and its wild relative, *Ipomoea lacunose*. Int Conf Sweetpotato Food Health Fut 583:81–85

Zhang Y, Liang Z, Zong Y, Wang Y, Liu J, Chen K, Qiu J, Gao C (2016) Efficient and transgene-free genome editing in wheat through transient expression of CRISPR/Cas9 DNA or RNA. Nat Commun 7:12617

Zhou C, Duarte T, Silvestre R, Rossel G, Mwanga RO, Khan A, George AW, Fei Z, Yencho GC, Ellis D, Coin LJ (2018) Insights into population structure of East African sweetpotato cultivars from hybrid assembly of chloroplast genomes. Gates Open Res 2:41

Zhu HC, Li C, Gao CX (2020) Applications of CRISPR-Cas in agriculture and plant biotechnology. Nat Rev Mol Cell Bio 21:661–677

Sweetpotato Breeding in the Genomic Age: Harnessing Databases, Bioinformatics, Digital Tools, and Genomic Insights

Bryan J. Ellerbrock, Christiano C. Simoes, Srikanth Kumar Karaikal, Christine M. Nyaga, and Lukas A. Mueller

Abstract

Various large breeding projects have been developed to modernize sweetpotato breeding with new ambitious targets and new genomic methods. As the data sources that these methodologies rely on grow in volume, so grows the importance of an efficient data management system. In this chapter, we introduce a digital breeding ecosystem centered around Sweetpotatobase, a database platform tailored for sweetpotato breeders across the globe to manage their breeding data effectively. We highlight a handful of complementary Android applications designed for data collection: Field Book, for phenotypic data, Coordinate, for genotypic data, and Intercross, for crossing data. We discuss the importance of roundtripping, and how BrAPI, a standard for breeding data transfer, can facilitate this via automated transfers. We present some of the many features of Sweetpotatobase that can be leveraged once the ecosystem is up and running—the Search Wizard, which explores and retrieves data intuitively and efficiently, the Pedigree Viewer, which can visualize allele inheritance patterns through pedigrees, the sweetpotato ontology, which defines standardized and measurable traits. Lastly, we discuss how Sweetpotatobase is empowered with analysis features, from a mixed model tool to genome-wide association studies (GWAS), principal component analysis (PCA), and stability analysis.

Keywords

Genome-based Breeding · Genomic selection · Breeding databases · Digital ecosystem · Breedbase

11.1 Introduction

Sweetpotato (*Ipomoea batatas*) is a widely consumed vegetatively propagated crop with particular importance as a subsistence crop in Africa. However, due to its hexaploidy ($2n = 90$), low flowering, and outcrossing phenotypes, it is a challenging crop for breeding (Campos and Caligari 2017). During the last decade, large breeding projects were implemented, including the SASHA, GT4SP, and

B. J. Ellerbrock
Clemson University, Clemson, SC, USA
e-mail: bryane@clemson.edu

C. C. Simoes · S. K. Karaikal · C. M. Nyaga · L. A. Mueller (✉)
Boyce Thompson Institute, Ithaca, NY, USA
e-mail: lam87@cornell.edu

S. K. Karaikal · C. M. Nyaga · L. A. Mueller
Cornell University, Ithaca, NY, USA

© The Author(s) 2025
G. C. Yencho et al. (eds.), *The Sweetpotato Genome*, Compendium of Plant Genomes,
https://doi.org/10.1007/978-3-031-65003-1_11

SWEETGAINS projects (Girard et al. 2017; Wu et al. 2018), intending to modernize sweetpotato breeding, address issues such as pathogen susceptibility, increase nutritional value, and to work toward the application of new methods, such as genomic selection.

A characteristic of these newer, genome-based breeding methods is their data-intensive nature. This characteristic prompted the projects to focus on enhancing the available infrastructure to handle the large-scale phenotyping and genotyping datasets that are required. Such infrastructure includes databases such as Sweetpotatobase (https://sweetpotatobase.org/), which was established based on the Breedbase software (https://breedbase.org/) (Morales et al. 2022). It implements a digital ecosystem that can facilitate the work of a breeding program aligned with these projects' goals. In addition, these projects addressed several other big obstacles, including the availability of the full genome sequence of sweetpotato (Chap. 2), the missing tools for analyzing polyploid genomes (Chaps. 4, 5) (Campos and Caligari 2017; Mollinari et al. 2020), and the creation and refinement of an appropriate ontology to describe the traits of sweetpotato (see https://cropontology.org/term/CO_331:ROOT).

In this chapter, we describe the available database infrastructure for sweetpotato. We also discuss how breeding programs can benefit from the available system and gain the maximum benefits by following established best practices for data management and workflows in the complex reality of a breeding program. Further resources are also available to get more information. For example, a good overview of the sweetpotato breeding community, best practices, new traits and methods, links to other resources, and many other aspects, is available from the Sweetpotato Knowledge Portal at https://www.sweetpotatoknowledge.org/.

Conventional breeding is a complex process, and the complexity is scaled upwards as genome-related data is integrated into breeding decisions. Importantly, some prerequisites must be satisfied before one can even start thinking about such an endeavor. The most important

prerequisites are: (1) A high-quality genome reference sequence that will facilitate the genotyping process; (2) algorithms that can be used to predict traits from genome data; (3) ontologies that describe the traits in the crop at hand, with well-defined data formats such as scales or categories. In a hexaploid system such as sweetpotato, (1) and (2) are much harder to achieve than for diploids because of the difficulty of assigning sequence reads to individual chromosomes. Using special techniques, the sequence of the diploid progenitors (Wu et al. 2018) and the hexaploid sweetpotato has only recently been completed (see Chap. 2). A complete database and website about the sequenced sweetpotato genomes is available at http://sweetpotato.uga.edu/, featuring interactive genome browsers and utilities such as BLAST searches. Polyploid genomes are much harder to genotype; while there are only three possible states in a diploid marker, the number of states increases rapidly with increasing ploidy levels. New methods had to be developed to identify the genotypes of hexaploid sweetpotato reliably, which is described in Chap. 4 (Campos and Caligari 2017; Mollinari et al. 2020). For the trait descriptors, work in collaboration with the crop ontology project (https://cropontology.org/) (Shrestha et al. 2012) yielded a standardized ontology (https://cropontology.org/term/CO_331:ROOT), which over the years has been improved and extended to adapt to the changing needs of the breeders, as new methods, such as near-infrared spectroscopy (NIRS), were introduced for many traits. New areas of interest, such as cooking quality, have been added to the phenotyping repertoire.

The fourth prerequisite, which is the focus of this chapter, is a strong commitment to strict data management principles and the establishment of the necessary data management infrastructure. It comprises a breeding database and data collection tools that seamlessly integrate into a digital breeding ecosystem, where data is collected and processed digitally, and breeding decisions can be made right in the database based on the latest data.

11.2 Digital Breeding Ecosystem

Breeding decisions are only as good at the data they are based upon. As selection strategies grow in sophistication, it is all the more important to ensure that the data underlying their complex models is inter-related, accessible, and free of preventable error. The key to accomplishing these goals is to produce, transfer, and store breeding data entirely within a digital ecosystem (Fig. 11.1).

At its surface, a digital ecosystem is composed of tools for data collection in the field or lab and for data analysis in the office. Underlying the tools is a programming interface that enables them to communicate automatically with the core of the ecosystem, a central database that stores, combines, and disperses the data. Passing data along from tool to database to tool for different steps of the breeding process without ever leaving the digital ecosystem is known as roundtripping. Roundtripping involves an initial setup cost to populate the database with breeding material to be tested and with traits to be measured. Then the activities of each breeding step can be tracked digitally. First, by creating database objects corresponding to the physical material, then by identifying the physical material with barcoded labels, and finally by collecting measurements on the physical material using the standardized traits. Once up and running the roundtripping process pays dividends as data is collected and related to each other with less effort, greater speed, and higher fidelity.

11.2.1 Data Collection

The data collection steps in the roundtripping process involve intricate workflows that need to be well-defined based on SOPs and can be very different for different crops. Data needs to be collected on the physical attributes of breeding lines (phenotypes), but also genetic data (genotypes) and relationship data (pedigrees). The following digital tools are used in sweetpotato programs for collecting each of these data types.

11.2.1.1 Phenotypic Data
There are many different approaches to collecting data in the field. Some examples include

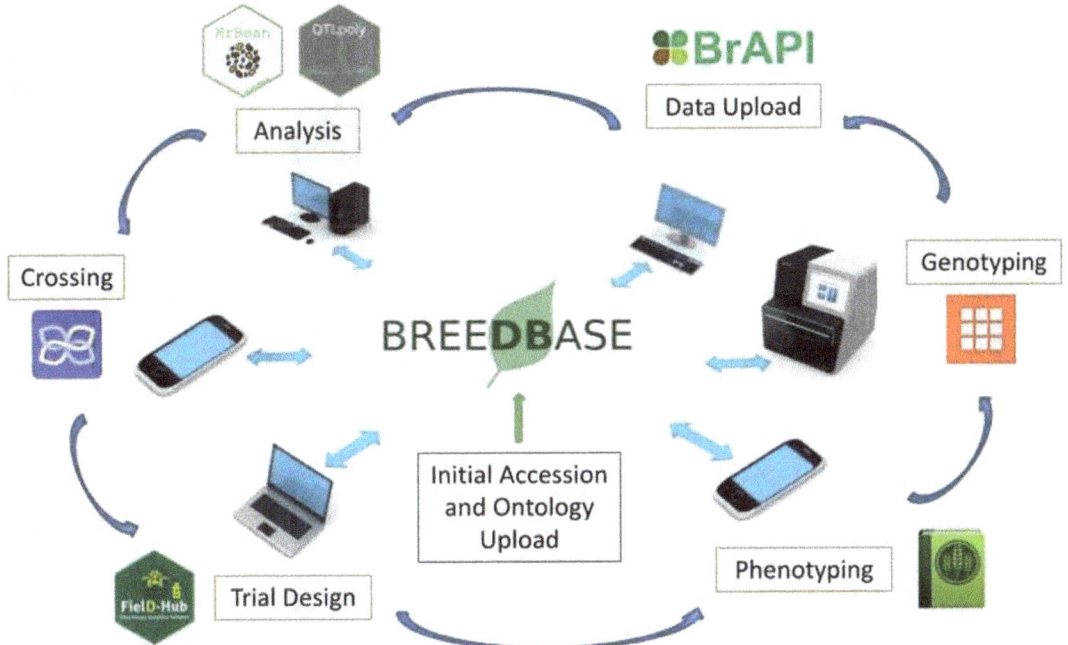

Fig. 11.1 Roundtripping within a digital breeding ecosystem

paper notebooks, digital spreadsheets, and custom breeding software on hand-held devices. Each has advantages and disadvantages, but we have found that Field Book app provides the ideal combination of features for most situations.

Field Book is an open-source Android app that can be used to collect data on plants in breeding and research applications (Rife and Poland 2014). Its data entry is efficient, eliminates the need for data transcription, and reduces the risk of errors. It runs on a wide range of inexpensive hardware, allowing consumer-grade technology to be used in environments where cost and inflexibility have been limiting factors.

An important consideration with Field Book is how one identifies the plot or plant that is being phenotyped in the field. Field Book provides a search interface to find the desired entry by plot or plant name, and it is possible to move automatically to the next entry in the field using on-screen buttons. However, we have found that barcoding the field is currently the best solution for routine identification of plots or plants. Field Book fully supports barcode scanning using the tablet camera, including QR codes, and Breedbase can generate PDFs for printing the labels. In addition, Field Book supports taking images in the field, which are automatically associated with the corresponding plot (Fig. 11.2).

For entering trait values, input screens adapt to the format of a trait; for example, for a categorical trait with categories 1, 2, 3, 4, and 5, five corresponding buttons will be displayed; it is impossible to enter an illegal value. A number pad is shown for numerical values, and for dates, a date selector, and so forth.

Roundtripping in the context of Field Book means that the trial layouts and traits have to be created or be available in the database and have to be exported to Field Book. Field Book essentially attaches data to the pre-existing data object identifiers such that the collected data can easily be uploaded back to the database.

Until recently, this data transfers from the database to Field Book and back involved file transfers, with the complexity of connecting the tablet to a computer, creating the necessary

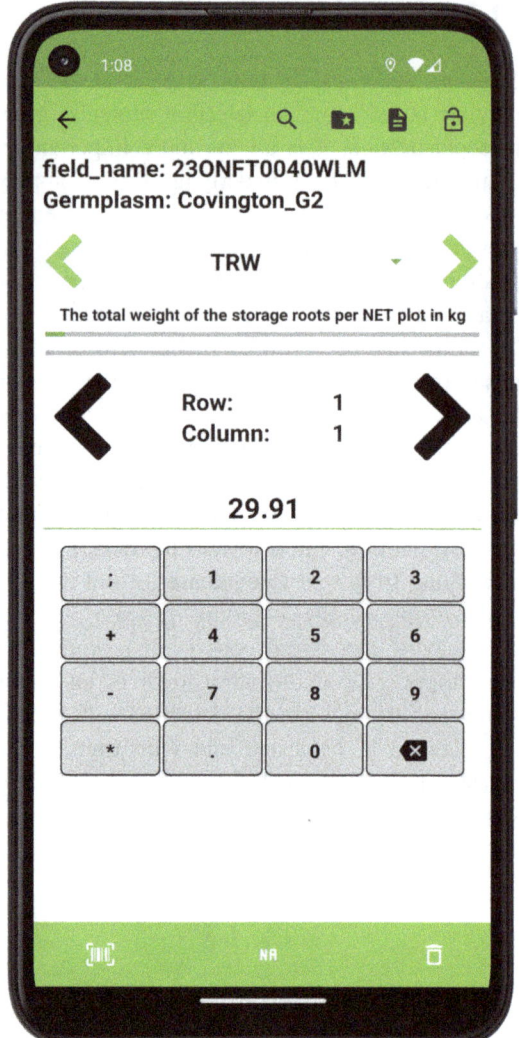

Fig. 11.2 Field book app's collect screen with collected sweetpotato data

files, and finding and transferring the files. Now, a BrAPI-based API (Selby et al. 2019) can be used, allowing Field Book to automatically import, export, and sync data from a BrAPI-enabled database over any internet connection with just the click of a button. This greatly reduces the amount of work required and makes the process of collecting data in parallel with multiple devices much easier.

Unfortunately Field Book is not a one-size fits all solution for phenotypic data types. Some data must be collected by incompatible hardware or in unsuitable workflows or conditions. In these cases, the roundtripping process can still be maintained by ensuring the necessary identifiers are propagated through the process using barcoded labels, or by making use of the standardized BrAPI calls.

A common example of one of these alternative workflows is the collection of near-infrared spectroscopy (NIRS) spectra. NIRS data collection requires the use of specialized benchtop or hand-held hardware, and outputs large quantities of spectral data. Regardless of the technology used, roundtripping can be maintained by propagating the unique identifiers of samples through the whole process so they are included in the output data. This ensures that the resulting spectra can be easily loaded and linked to the proper objects within Sweetpotatobase.

11.2.1.2 Genotypic Data

Genotypic data is a complex data type, and tissue sample collection, processing, and analysis can be challenging. Collection of samples in the field is often an error-prone manual process; a breeder must go to the field, collect and label individual samples, and lay those samples out on a plate that can be submitted for sequencing, taking particular care to prevent mix ups so that when data is returned from a genotyping facility, it can easily be connected back to the original samples (Fig. 11.3).

In the digital ecosystem, the Coordinate app provides support for these activities. Coordinate is a flexible, open-source Android app that is used to collect and organize samples. Coordinate functions by defining templates and then collecting data in grids created from those templates. The plot or plant barcode can be scanned with the app to identify the samples collected and a unique identifier is generated for each sample. The samples are arrayed in 96 well plates, and the corresponding data is uploaded to the database, which can, for some providers, submit the data automatically to genotyping facilities.

In turn, Sweetpotatobase can serve as a repository for various genetic markers that are output by the genotyping process. Supported types include Single Nucleotide Polymorphism (SNP)

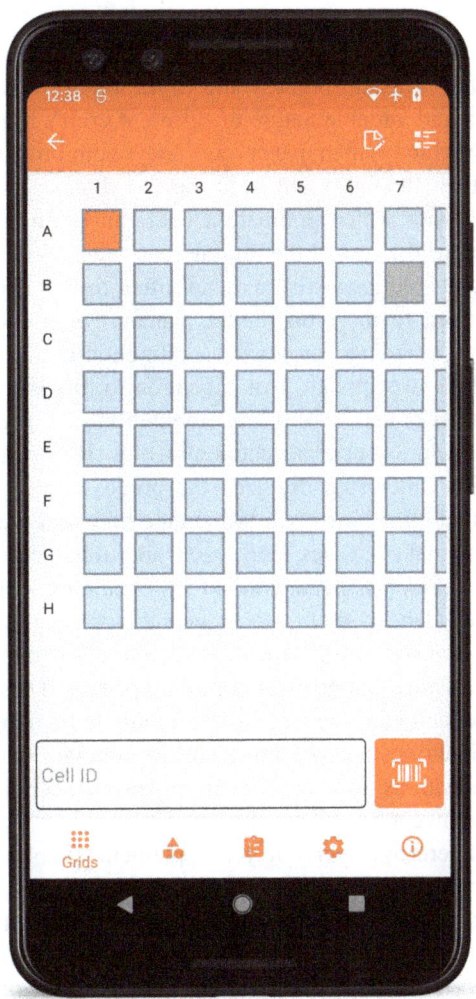

Fig. 11.3 Coordinate data collection screen

markers, Kompetitive Allele-Specific PCR (KASP) markers, and Simple Sequence Repeats (SSR) (Morales et al. 2020). By associating these genotypes back to the source tissue samples generated by Breedbase and tracked in Coordinate, they are automatically integrated with the broader set of phenotypic and relationship data. This contextualized genotypic data is instrumental in unraveling the genetic diversity of sweetpotato. It enables sweetpotato breeding programs to deploy a powerful set of tools, from purely quantitative genomic selection (GS) models to genome-wide association studies (GWAS) that identify specific inheritance patterns and marker-assisted selection (MAS) strategies that exploit them.

11.2.1.3 Crossing Data

Crossing is frequently the least digitzied part of the breeding process due to the highly complex and variable nature of different crop's biology. Handwritten paper tags are common due to their flexibility but limit the speed and accuracy with which pollination data can be linked back to the rest of the digital ecosystem. Where digital solutions exist, they are often tightly customized to the crop (btract, banana) or a specific breeding program (pollination-toolbox, NCSU sweetpotato). An exception to this is the Intercross app, a beta implementation of a general solution to cross data collection (Rife et al. 2022). Intercross digitizes the process through a streamlined interface to manage potential parents, make crosses, and track additional cross data. Parents are identified by scanning barcoded labels while newly created pollinations are tracked using labels produced on-demand by a Bluetooth-connected mobile printer. These barcoded labels ensure fast and reliable tracking of identifies and are important to maintain data connections in downstream processes such as seed inventory and seedling trials.

Intercross does not yet implement all of its imports and exports via BrAPI, but the necessary file formats are interoperable with Sweetpotatobase. The Sweetpotatobase crossing experiment page can generate both a parent file to import the necessary male and female ids, as well as a wishlist file used to set pollination targets for specific parental combinations.

Exported crossing data can be updated in a standard spreadsheet format accepted by Sweetpotatobase. When uploaded, this data automatically populates details including unique pollination event ids, timestamps, operator name, and optional fields like flower number. As the seed generated is collected, planted, and selected, any downstream data collected can be linked back to the original cross using the unique id encoded in the barcoded label. Digital tracking of this data such as seed numbers and progeny names allows Sweetpotatobase to automatically calculate pollination success rates, selection percentages, and to populate the pedigrees of newly selected material (Fig. 11.4).

Fig. 11.4 Details of a sweetpotato pollination recorded in the Intercross app

11.2.2 Data Transfer

As with other technologies, breeding data transfer has come a long way, from transcription, to manual transfer of digital files, to Breeding API (BrAPI) calls (Selby et al. 2019). While relatively new and still under active development, all the digital tools we recommend here have adopted the BRAPI standard as a way to automate data transfer between freestanding software tools and into flexible analysis environments like RStudio. These automated transfers are

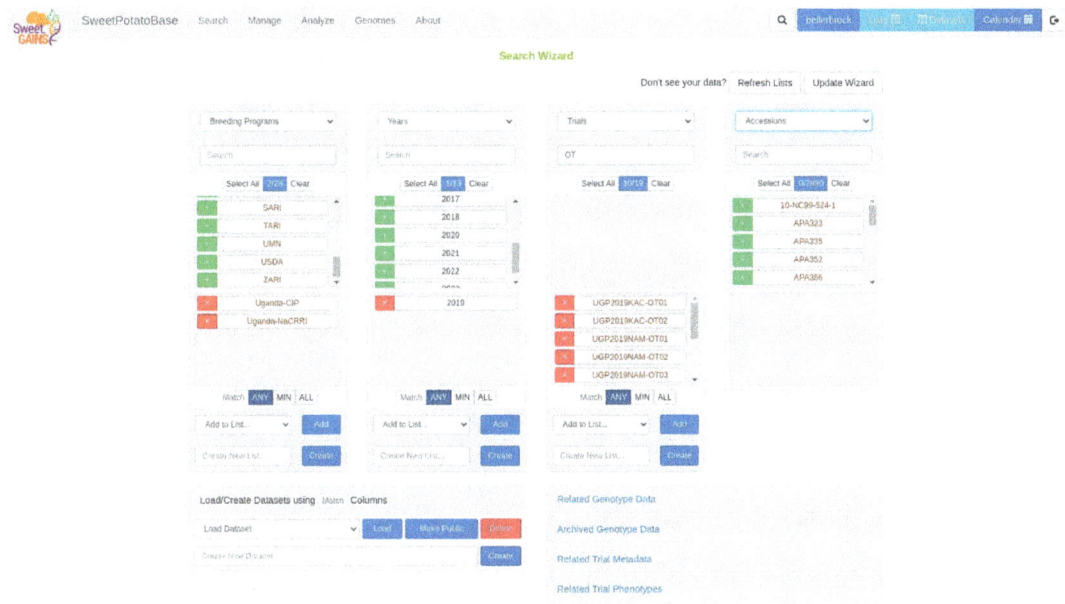

Fig. 11.5 Slice of sweetpotato data in the Search Wizard

the glue that keep the ecosystem together and can be the difference between a nearly friction-less roundtripping experience, or a tedious file-transfer process that limits the process or breaks it entirely. While the digital ecosystem is flexible in that the available tools are optional and may change over time, it is crucial that any additions to the system speak the shared language.

11.2.3 Data Management

11.2.3.1 The Search Wizard

In Sweetpotatobase, as in other Breedbase data-bases, the Search Wizard is a major query tool, empowering users with an intuitive and efficient approach to data exploration and retrieval. In the Search Wizard, the data in the database is viewed as a multi-dimensional cube, in which the dimensions represent attributes of the trial data, such as location, year, breeding program, accessions, traits, and so forth. The Search Wizards allows the specification of data items along these dimensions to create intersects in the

data cube, efficiently generating highly precise datasets that can be stored in the database and used for downstream analyses (Fig. 11.5).

Analysis tools that support data input by wiz-ard datasets include heritability assessments, stability evaluations, genomic selection method-ologies, and mixed models.

A significant advantage lies in the tool's capacity to combine these different dimensions and store each parameter individually, both within lists and datasets. Lists prepared through the Search Wizard are automatically validated, whereas manually created lists require addi-tional validation.

Lists and datasets play an indispensable role in many functionalities within the Breedbase platform and expedite activities from trial crea-tion to seamlessly integrating accessions and facilitating the use of various other tools.

11.2.3.2 The Pedigree Viewer

Pedigrees assist breeders in identifying cross-breeding combinations that result in desirable traits, such as high yields and resistance to biotic and abiotic stress. Breeders can observe allele inheritance through pedigrees and understand their influence on trait expression and breeding strategies. Breeders can generate large pedigree structures using this genetic and phenotypic data. This helps breeders and researchers to make informed decisions about which plant lines to use in subsequent crossings.

Sweetpotatobase contains over 150,000 *Ipomoea batatas* accessions, encompassing a wealth of genetic diversity sourced across multiple breeding programs. Accessions can be linked to pedigrees, which are generated when accessions are crossed using the crossing tool or can be uploaded into the system via a specific table format (Fig. 11.6).

Within Sweetpotatobase, pedigrees can be displayed and interacted with using the Pedigree Viewer. By default, the viewer shows an accession male and female parents, identified using color-coded lines. Purple arrows indicate nodes that can be expanded to display more relationships, either additional progeny, siblings, or parents, depending on the location and direction of the arrow. Access to pedigrees in this

way adds depth and lineage context to the stored germplasm and provides valuable insights into genetic relationships and traits.

11.2.3.3 The Trait Ontology

The trait ontology is an important aspect of any breeding program, as it defines which traits can be measured and stored in the database, and it also standardizes how a trait is measured to ensure comparability of the results, potentially across breeding programs worldwide (Shrestha et al. 2012). The sweetpotato ontology in the database consists of 327 variables, which refer to traits with methods and scales (Fig. 11.7).

The Breedbase system allows post-composing of trait variables using other orthogonal ontologies, which can specify sampling conditions, temporal components, or sample treatments. These can be formed on the fly, while traits in the ontology itself can only be changed through a request to the ontology team to ensure standardization and avoid duplication of terms. Post-composing increases the flexibility of the annotations without sacrificing the

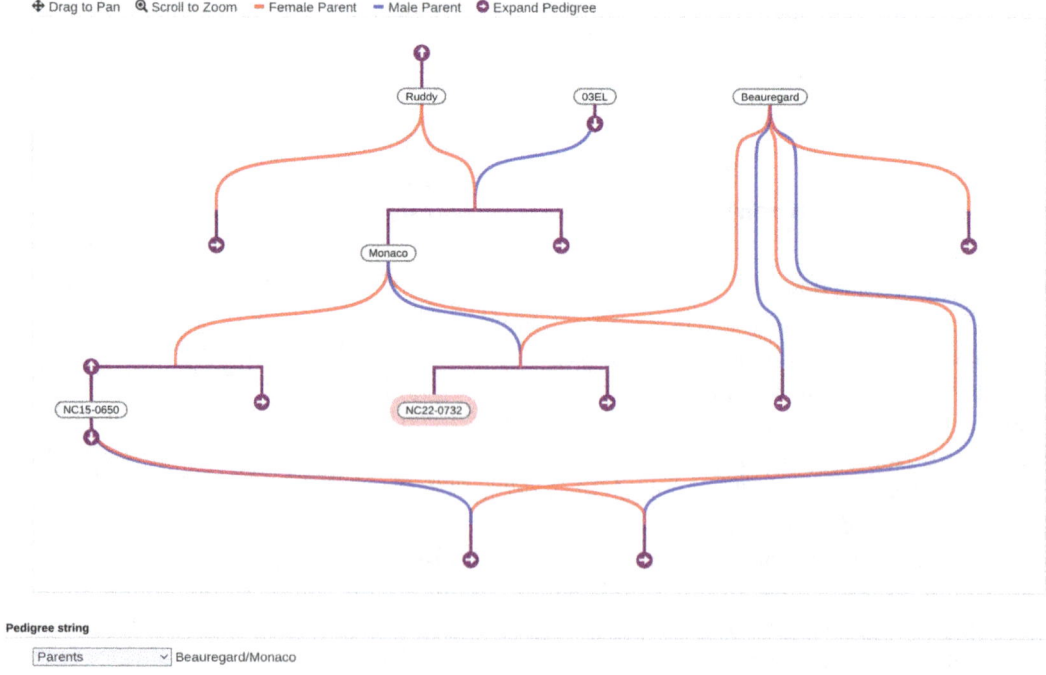

Fig. 11.6 Sweetpotato accession's pedigree, visualized using the Pedigree Viewer

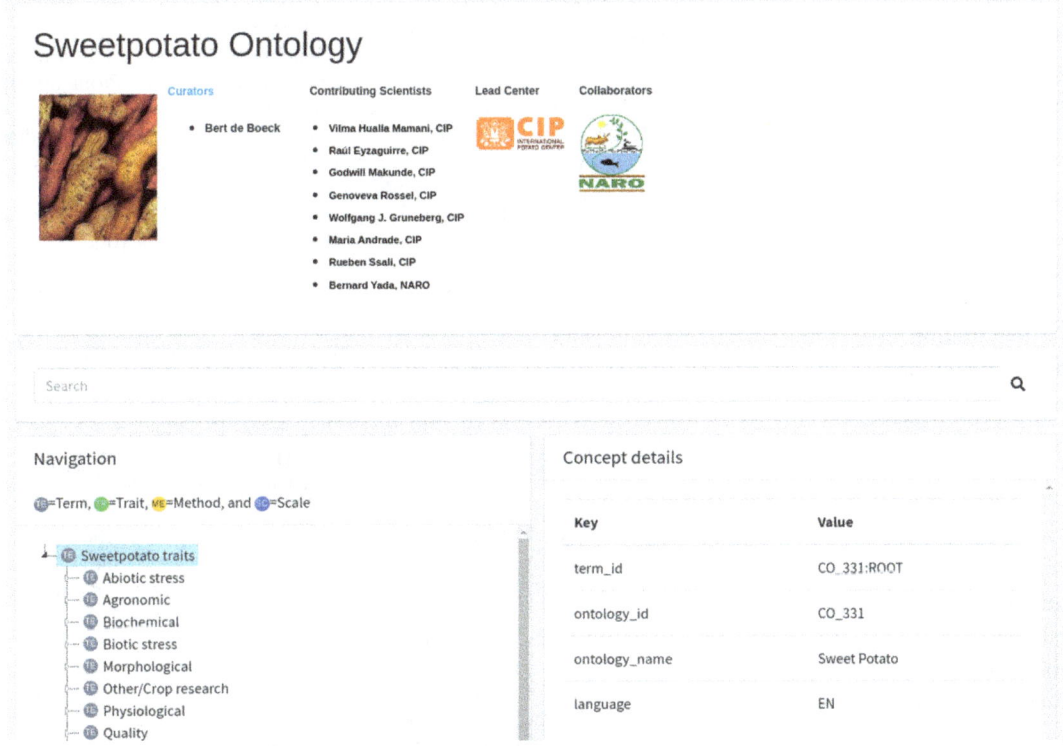

Fig. 11.7 Sweetpotato ontology, as seen at cropontology.org

standardization of the terms. In Sweetpotatobase, 105 post-composed traits have been created by users, showing the platform's adaptability and usability in storing diverse data and enhancing its accessibility for researchers and breeders alike. The sweetpotato ontology has been continually developed in collaboration with the crop ontology project (https://cropontology.org/).

11.2.3.4 Data Analysis

Analytical tools within Sweetpotatobase offer a comprehensive suite for sweetpotato breeders, empowering users with nuanced insights into their crop's performance. Available phenotypic analyses include analysis of variance (ANOVA) to assess the significance of comparisons of sweetpotato traits across different varieties or treatments. Heritability and stability analysis help quantify the extent to which observed traits are genetically inherited and how consistent they remain across environments. This insight aids breeders in selecting superior sweetpotato

cultivars with desirable and stable characteristics. The GWAS tool (Morales et al. 2020) runs GWAS analyses right in the database, using datasets selected by the Wizard tool. Population structure analysis elucidates genetic diversity and relatedness among sweetpotato accessions, informing breeding strategies for maximizing heterosis and minimizing inbreeding depression. Mixed models complement this by accounting for genetic relatedness in statistical analyses, offering more accurate trait predictions and breeding value estimations. A built-in genomic selection workflow called solGS (Tecle et al. 2014) can be used to run the entire genomic selection pipeline in the Breedbase system, including the generation of models and prediction of GEBVs from genotypic information.

The built-in support for BrAPI in Breedbase means that BrAPI enables analysis tools that can be used with Sweetpotatobase. One such tool is the MrBean analysis tools offer a user-friendly platform tailored for molecular marker data analysis, particularly suited for plant breeding and genetic studies. Its intuitive interface streamlines tasks like marker-trait association studies, population structure analysis, and genomic prediction (Aparicio et al. 2024).

Using MrBean and other BrAPI-enabled tools, researchers can complement the built-in function of Breedbase to conduct comprehensive phenotypic and genetic analyses. Its functionality extends to diversity analysis, facilitating the exploration of genetic variation within and among populations.

11.3   Conclusion

Sweetpotatobase provides a digital ecosystem to assist breeders worldwide with managing sweetpotato breeding data and making more informed, data-based breeding decisions. Sweetpotatobase integrates many different data types, such as phenotypic and genotypic information, and associated analysis tools that can aid in improving selections. Contributing data into one system by many breeding programs enables better collaboration and builds larger models with more predictive power, increasing genetic gain. The database seamlessly interfaces with digital data acquisition tools from the PhenoApps project, including the Field Book, Coordinate, and Intercross apps. It supports the BrAPI standard for interfacing with tools and analyses. A comprehensive trait vocabulary (ontology) has been developed, constantly updated, and is used routinely in more than a dozen registered breeding programs. Databases such as Sweetpotatobase are part of the necessary foundation for any modern breeding program.

References

Aparicio J, Gezan SA, Ariza-Suarez D, Raatz B, Diaz S (2024) Mr Bean: a comprehensive statistical and visualization application for modeling agricultural field trials data. Front Plant Sci 14:1290078

Campos H, Caligari PDS (2017) Genetic improvement of tropical crops. Springer, New York

Girard AW, Grant F, Watkinson M, Okuku HS, Wanjala R, Cole D, Levin C, Low J (2017) Promotion of orange-fleshed sweet potato increased vitamin A intakes and reduced the odds of low retinol-binding protein among postpartum Kenyan women. J Nutr 147:955–963

Meier L (2022) ANOVA and mixed models: a short introduction using R. CRC Press, New York

Mollinari M, Olukolu BA, da Pereira G, S, Khan A, Gemenet D, Yencho GC, Zeng Z-B, (2020) Unraveling the hexaploid sweetpotato inheritance using ultra-dense multilocus mapping. G3 10:281–292

Morales N, Bauchet GJ, Tantikanjana T, Powell AF, Ellerbrock BJ, Tecle IY, Mueller LA (2020) High density genotype storage for plant breeding in the Chado schema of Breedbase. PLoS ONE 15:e0240059

Morales N, Ogbonna AC, Ellerbrock BJ, Bauchet GJ, Tantikanjana T, Tecle IY, Powell AF, Lyon D, Menda N, Simoes CC et al (2022) Breedbase: a digital ecosystem for modern plant breeding. G3 12:78. https://doi.org/10.1093/g3journal/jkac078

Rife TW, Poland JA (2014) Field book: an open-source application for field data collection on android. Crop Sci 54:1624–1627

Rife TW, Courtney C, Bauchet G, Neilsen M, Poland JA (2022) Intercross: an Android app for plant breeding and genetics cross management. Crop Sci 62:820–824

Selby P, Abbeloos R, Backlund JE, Basterrechea Salido M, Bauchet G, Benites-Alfaro OE, Birkett C, Calaminos VC, Carceller P, Cornut G et al (2019) BrAPI-an application programming interface for plant breeding applications. Bioinformatics 35:4147–4155

Shrestha R, Matteis L, Skofic M, Portugal A, McLaren G, Hyman G, Arnaud E (2012) Bridging the phenotypic and genetic data useful for integrated breeding through a data annotation using the crop ontology developed by the crop communities of practice. Front Physiol 3:326

Tecle IY, Edwards JD, Menda N, Egesi C, Rabbi IY, Kulakow P, Kawuki R, Jannink J-L, Mueller LA (2014) SolGS: a web-based tool for genomic selection. BMC Bioinform 15:398

Wu S, Lau KH, Cao Q, Hamilton JP, Sun H, Zhou C, Eserman L, Gemenet DC, Olukolu BA, Wang H, Crisovan E, Godden GT, Jiao C, Wang X, Kitavi M et al (2018) Genome sequences of two diploid wild relatives of cultivated sweetpotato reveal targets for genetic improvement. Nat Commun 9:4580

The Future of Crop Improvement in Sweetpotato: Merging Traditional and Genomic-Assisted Breeding Methods

12

Bonny Michael Oloka, Carla Cristina da Silva, Camila Ferreira Azevedo, Innocent Vulou Unzimai, Benard Yada, Wolfgang Grüneberg, Maria Andrade, Kenneth V. Pecota, Guilherme da Silva Pereira and G. Craig Yencho

Abstract

Crop improvement in sweetpotato has progressed slowly in many parts of the world largely due to its significant genetic complexity arising from its large autohexaploid genome, high heterozygosity, and self and cross-incompatibilities. New breeding tools have been developed to better understand this crop and its important agronomic and culinary traits. These tools and their application are reviewed here, and the path forward has been proposed. By incorporating these new genomic tools into breeding programs routinely alongside the traditional methods, crop improvement can be accelerated, leading to the delivery of clones with better genetics to farmers more quickly. This integration of genomics could propel sweetpotato into a new era, ultimately enhancing its productivity and profitability, which is crucial given the growing global population.

Keywords

Ipomoea batatas · Genome sequencing · Molecular markers · Linkage mapping · Quantitative trait loci · Marker-assisted selection

B. M. Oloka (✉) · K. V. Pecota · G. C. Yencho
Department of Horticultural Science, NC State University, Raleigh, NC 27695, USA
e-mail: boloka@ncsu.edu

G. C. Yencho
e-mail: craig_yencho@ncsu.edu

C. C. da Silva · C. F. Azevedo · I. V. Unzimai · G. da Silva Pereira
Department of Agronomy, Federal University of Viçosa, Viçosa, Brazil

B. Yada
National Crops Resources Research Institute, National Agricultural Research Organization, Kampala, Uganda

W. Grüneberg
International Potato Center, Lima, Peru

M. Andrade
International Potato Center, Maputo, Mozambique

12.1 Introduction

Crop production continues to face challenges from a rapidly growing human population, climate change, and weather variability putting pressure on land and water resources. Advances in plant breeding technologies have in the past produced crops that adapt to biological and physical stresses much faster than they appear (Montesinos-López et al. 2018; Pipitpukdee et al.

G. C. Yencho et al. (eds.), *The Sweetpotato Genome*, Compendium of Plant Genomes, https://doi.org/10.1007/978-3-031-65003-1_12

2020). However, the time required to achieve these genetic gains and feed the world has remained constant for several years. To increase agricultural productivity growth especially for sweetpotato, a substantial investment in innovation, adoption, use, and better identification of the most appropriate technologies and practices for improved performance is required. Increased adoption of improved varieties of sweetpotato is dependent upon increased productivity of the crop. Various biotic and abiotic constraints that affect sweetpotato productivity include sweetpotato virus disease (SPVD), *Alternaria bataticola* blight, sweetpotato weevils, inadequate and variable rainfall, and low soil fertility (Mwanga et al. 2021b). Other factors that improve adoption include consumer/market preferences and gender (Mwanga et al. 2021a), as well as a well-structured seed system for the crop.

Previously, successfully bred sweetpotato varieties were identified from thousands (50,000–100,000) of seedlings germinated from multi-parent crossing block nurseries (Fig. 12.1). These seedlings would then undergo a series of field trials and testing for several years and seasons which typically lasted eight years. In the early stages of trialing, only a few key traits would be measured because of the many numbers of individuals. In the later stages, a greater number of traits (approximately 40) would be measured on the remaining few high-performance individuals in on-farm and national performance trials. Suitable parents for the next breeding cycle would be identified at this point. In the Eastern African regions, which are a hotspot for SPVD, plants would escape infection in the early stages of testing only to show up a few years later thus complicating efforts to release a truly resistant variety that has acceptable qualities for all the other traits of importance.

In comparison to other major food crops globally, sweetpotato is regarded as the world's seventh most important crop species. Yet it has only received modest funding to develop its genomic breeding resources. Recent efforts through the genomic tools for sweetpotato improvement (GT4SP) and the genetic advances and innovative seed systems for sweetpotato

(SweetGAINS) projects, both supported by Bill and Melinda Gates Foundation (BMGF), have delivered a suite of genomic tools for improvement of sweetpotato breeding. These tools include two fully sequenced and annotated wild diploid *I. trifida* and *I. triloba* reference genomes (Wu et al. 2018); a genome browser, and a sequenced sweetpotato *I. batatas* hexaploid reference genome (http://sweetpotato.uga.edu/); a sequencing-based genotyping platform for highly heterozygous hexaploid sweetpotato, OmeSeq Qrrs—derived from GBSpoly (Wadl et al. 2018), with supporting bioinformatics tools; three high-density genetic maps of hexaploid sweetpotato—for the BT (Mollinari et al. 2020), TB (Amankwaah 2019) and NKB (Oloka 2018) mapping populations; dosage-dependent SNP calling, phasing, and linkage mapping algorithms for autopolyploids—MAPpoly R package (Mollinari et al. 2020); robust polyploid QTL analysis—QTLploy R package (Da Silva et al. 2020); and a breeding program database—sweetpotato base for the management of breeding data (Morales et al. 2022). Altogether, these tools have been very useful in the sweetpotato breeding community in understanding the genetic architecture of key traits and their underlying molecular mechanisms (Gemenet et al. 2020a; Oloka et al. 2021).

With the available genomic resources on hand, the sweetpotato breeding community is now able to utilize breeding values and marker information to make selections and advancement decisions. We are also working on the proof-of-concept for genomic selection in sweetpotato and have developed three training populations from three breeding programs: (1) the National Crops Resources Research Institute (NaCRRI) population consists of 324 individuals with a primary focus on sweetpotato weevils; (2) the International Potato Center (CIP) population consists of 1200 individuals focusing on sweetpotato virus disease; and (3) the North Carolina State University (NCSU) population consists of 504 individuals focusing on guava and southern root-know nematodes. These populations were phenotyped extensively and genotyped to be used for early parent identification and product

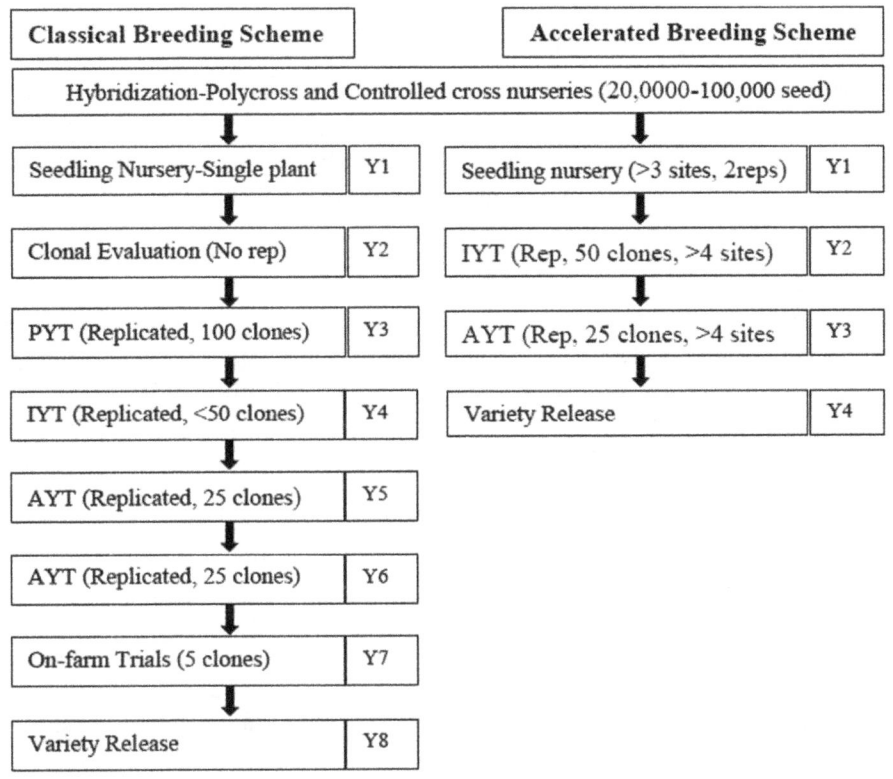

Classical Breeding Scheme		Accelerated Breeding Scheme	
Hybridization-Polycross and Controlled cross nurseries (20,0000-100,000 seed)			
Seedling Nursery-Single plant	Y1	Seedling nursery (>3 sites, 2reps)	Y1
Clonal Evaluation (No rep)	Y2	IYT (Rep, 50 clones, >4 sites)	Y2
PYT (Replicated, 100 clones)	Y3	AYT (Rep, 25 clones, >4 sites	Y3
IYT (Replicated, <50 clones)	Y4	Variety Release	Y4
AYT (Replicated, 25 clones)	Y5		
AYT (Replicated, 25 clones)	Y6		
On-farm Trials (5 clones)	Y7		
Variety Release	Y8		

Fig. 12.1 Classical and accelerated breeding scheme of sweetpotato (© B. Yada 2014, Dissertation, NCSU)

development. The vast amounts of data generated can be compared between breeding programs because standard operating procedures were embraced early on that include a standard crop ontology (www.cropontology.org) for trait nomenclature, a breeding program database (www.sweetpotatobase.org) for data management and curation, and electronic data capture using fieldbook app.

The most expensive and time-consuming component of sweetpotato crop improvement is trialing and phenotyping thousands of clones, especially in early stages of breeding. We believe that genetic gain can be achieved faster with genomic-assisted breeding as useful clones are identified earlier on in the breeding program without going through many cycles of trialing (Heffner et al. 2009). Genomic selection has been successfully implemented in a number of crop improvement programs including maize (Bernardo and Yu 2007), cassava (Ozimati et al.

2019), soybean (Duhnen et al. 2017), and eucalyptus (Resende et al. 2012) among several other crops. Exploiting the full potential of sweetpotato world over requires addressing the root causes of yield gaps as well as merging proven genomics-assisted breeding approaches with traditional breeding methods in the crop.

This book highlights where we have come from in sweetpotato crop improvement and our efforts to merge classical breeding approaches to genomics-assisted breeding in applied breeding programs that actively release sweetpotato varieties. Strong and effective breeding programs require good leadership, breeding resources that include a stellar team, significant time, and consistent funding. With these in place, very good returns on investment will be realized as improved varieties that benefit farmers, societies and the environment will be released with high variety turnover and adoption (Mwanga et al. 2021b). We further demonstrate the importance of identifying, training, and bringing together teams to address common problems in food systems. We highlight challenges, lessons learned, and how we have approached different problems encountered as we dissect the complex genome of sweetpotato to improve the traits dearest to sweetpotato stakeholders.

12.2 Genetic Improvement of Sweetpotato Traits

There are numerous traits in the sweetpotato ontology (www.cropontology.org), all of which need to come together perfectly to make the ideal variety. However, some are controlled by single genes and are relatively easier to breed due to their high heritability. In this section, we will concentrate on our efforts to improve the basic traits which are 'must-have' traits in breeding programs. These include resistance to sweetpotato weevils, sweetpotato virus disease, and root-knot nematodes. We will also look at 'value-added' traits like β-carotene and sugars, and dive deep into how genomic tools are being used to improve them.

12.2.1 Breeding for Resistance to Sweetpotato Weevils

The sweetpotato weevil, *Cylas* spp., is the most serious insect pest of sweetpotato worldwide. *Cylas puncticollis* (Boheman) (Fig. 12.2a) and *Cylas brunneus* (Fabricius) (Fig. 12.2b) are uniquely African species (Downham et al. 2001). The larvae feed on roots but are not always readily observed until they have caused significant damage. Adults are also difficult to detect given their nocturnal habitat. 'New Kawogo,' a 1995 released landrace cultivar in Uganda (Mwanga et al. 2001), is reported to be resistant to the notorious weevil species *C. puncticolis* and *C. brunneus*. The mechanism of resistance is active and is associated with the high concentrations of six hydroxinnamic acid (HCA) esters mainly on the surface of the roots (Stevenson et al. 2009). These esters were identified as

hexadecylcaffeic acid, hexadecylcoumaric acid, heptadecylcaffeic acid, octadecylcoumaric acid, and 5–0-caffeoylquinic acid (Stevenson et al. 2009; Anyanga et al. 2013; Yada et al. 2017a). Host plant resistance provides an effective and long-lasting component of any integrated pest management program. However, the development of high-yielding, commercially acceptable weevil-resistant varieties has not been successful over the years due to the lack of heritable resistance in the existing sweetpotato germplasm pool (Anyanga et al. 2017).

A number of recently published studies have contributed much to our understanding of the genetic and biochemical basis of resistance to SPW observed in 'New Kawogo.' Yada et al. (2015) identified 12 SSR markers that were associated with SPW resistance in a 287-clone segregating population derived from a biparental cross between 'New Kawogo' and 'Beauregard.' Thereafter, Anyanga et al. (2017) used the same population to improve our understanding of the biochemical basis of resistance observed in 'New Kawogo' and reported the segregation of resistance conferred by hydroxycinnamic acid esters that occur on the surface of storage roots. This work was followed up by Oloka (2018) who developed an integrated genetic linkage map of sweetpotato in the NKB population using single nucleotide polymorphism (SNP) markers. In this work, they were not able to identify any significant QTL for SPW resistance. This was due to a number of factors, some of which they pointed out as insufficient population size, highly heterogeneous weevil population in the different environments where the trials were conducted, and significant genotype-by-environmental effects. Consequently, previous efforts have had limited utility for marker-assisted selection but have provided the foundation for further studies utilizing whole genome markers and robust bioassay phenotyping for SPW. For such a complex trait that requires expensive and laborious phenotyping to address, the need to tackle it using genomic-assisted breeding approaches has never been greater.

For genomic selection (GS) to deliver enhanced genetic gains for the traits of interest,

Fig. 12.2 Sweetpotato weevils *Cylus puncticollis* and *C. brunneus* as seen with the naked eye

we noted that many aspects of breeding operations had to be redesigned, with the emphasis on accurate phenotyping. Previously, SPW was phenotyped in the field using incidence and severity scores on harvested storage roots 130–150 days after planting. The SPW incidence score is simply the percentage of infected roots in the plot whereas the severity score is a 1–9 scale, where, 1 = no weevil damage on any root and 9 = severe damage symptoms on all roots in the plot (Grüneberg et al. 2019). After harvesting and data collection, storage roots are sampled and brought to the lab for choice and no-choice bioassay experiments (Nottingham et al. 1989). In these bioassays, individual roots are artificially infested with a given number of 10-week-old gravid female weevils and then observed for weevil feeding and oviposition. The number of adults that emerge after the infestation period is also counted and recorded. Research has shown that there is a positive correlation between sweetpotato weevil bioassays and field root infestation (Anyanga et al. 2017).

12.2.2 Resistance to Sweetpotato Virus Disease

Sweetpotato virus disease (SPVD) is a complex synergistic interaction of whitefly-transmitted *sweetpotato chlorotic stunt virus* (*Crinivirus*) and aphid-transmitted *sweetpotato feathery mottle virus* (*Potyvirus*). Resistance to SPVD has been identified as one of the main biotic stress breeding objectives for sweetpotato breeding programs in SSA as it causes yield losses of 50–90% in many high-yielding susceptible genotypes (Clark et al. 2012). Plants infected by SPVD can be easily recognized visually by growers due to their clear field symptoms which include stunting, chlorosis, mosaics, leaf narrowing, and distortion (Gibson et al. 1998; Clark et al. 2012) (Fig. 12.3).

The main management approach in regions where there is high virus pressure has been removal of visually infected plants from the field and use of clean planting materials (Aritua et al. 2007). This approach is effective in developed parts of the world but not in SSA where there is no formal seed system to ensure adequate and timely supply of clean seed (Tairo et al. 2005). Conventional breeding has not been successful over the years due to the limited sources of resistance in the available germplasm pool. There are few demonstrated sources of resistance present in released local landraces in SSA like 'New Kawogo' and 'Tanzania' as well as some wild relatives of sweetpotato (Karyeija et al. 2000). However, field screening for virus resistance is both slow and inefficient due to a number of factors. Vector populations fluctuate over seasons and years (Aritua et al. 2007) thus complicating trials to identify agronomically superior genotypes from large (50,000–100,000) populations of F_1 clones (Mwanga et al. 2021b). This results in plants escaping infection, only to show severe symptoms of SPVD after three or more years of planting. It may, therefore, not be ideal to use symptom severity as a selection criterion for SPVD in a segregating population (Clark et al. 2012).

Fig. 12.3 Sweetpotato plant (circled) showing severe symptoms of SPVD infection in the same plot with visually symptomless plants

A limited number of molecular studies have been conducted to improve our understanding of the resistance of sweetpotato clones to SPVD. The first studies were conducted by Mwanga et al. (2002), who identified two recessive genes, *spfmv1* and *spcsv1* from a biparental cross between 'Tanzania' and 'Bikilamaliya.' Amplified fragment length polymorphism (AFLP) and random amplified polymorphic DNA (RAPD) markers were used in this population to identify the two markers. Thereafter, AFLP markers were associated with SPVD resistance using discriminant analysis and logistic regression (Mcharo et al. 2005; Miano et al. 2008). However, the fact that AFLP and RAPD markers are dominant in nature has limited their utility in applied sweetpotato breeding programs. Regression analysis was used by Yada et al. (2017b) to associate simple sequence repeat (SSR) markers to SPVD resistance in a 'New Kawogo' by 'Beauregard' (NKB) biparental population. They identified seven SSR markers that were associated with resistance to SPVD in the NKB population, but the utility of these markers in other sweetpotato breeding populations has not been realized. It is important to find markers that have no ascertainment bias and are useful across breeding populations outside the study population.

12.2.3 The Root-Knot Nematode

Plant parasitic nematodes are major pathogens of many agricultural commodities globally (Agrios 2004). Root-knot nematodes (RKNs–*Meloidogyne* spp.) parasitize almost every species of vascular plants (Jones et al. 2013) resulting in billions of dollars in annual crop losses (Sasser 1980). In SSA, particularly Eastern Africa, RKNs affect a range of staple RTB crops including banana, cassava, and sweetpotato (Coyne et al. 2006; Karuri et al. 2017; Akinsanya et al. 2020). Much of the damage by RKNs in SSA goes undetected mainly due to their associations in disease complexes with fungi and bacteria. Infection by RKNs causes cracking of roots and secondary infections reducing the market value of root crops by directly affecting their quality (Oloka et al. 2021). Storage root cracking can also result from late-season rains after a long dry period and produce symptoms slightly similar to those caused by RKN infection. For this reason, storage root cracking is not recommended for use as a standard in management schemes for RKNs in sweetpotato.

In infected fields, neurotoxic nematicides alongside cultural practices have often been used as control strategies for the management of RKNs. Besides their prohibitive cost to small-scale growers of sweetpotato, the health and environmental risk posed by nematicides are great (Chitwood 2003). The use of resistant plant genotypes has always been the safest, most sustainable, and economic control strategy of plant parasitic nematodes. However, the effectiveness of crop improvement efforts depends on breeding program resources, the nematode environment, availability of efficient screening procedures, availability and identification of usable sources of durable resistance, and knowledge of the genetics and inheritance of resistance. To a great extent, these factors determine the breeding method and success of the research. Most widely grown and popular sweetpotato varieties globally are highly susceptible to plant parasitic nematodes (Oloka et al. 2021). In the USA, resistance has been identified in un-adapted clones whereas in SSA, resistance is found in low-yielding landraces with low nutritional value. Integrating this resistance in high-yielding adapted elite clones is the challenge before sweetpotato crop improvement programs.

Sweetpotato is a highly heterozygous autohexaploid crop with large genome and complex genetics. Its high heterozygosity means that when resistance is identified in an un-adapted clone, simple backcrossing to recover elite traits is not possible as this will unmask numerous unwanted lethal genes that are detrimental to the survival of the crop. In order to fast track the breeding process in the improvement of sweetpotato plants to plant parasitic nematodes, traditional breeding needs to be merged with genomic-assisted breeding technologies as well as innovative cultural management practices. To this end, a number of sweetpotato breeding programs have developed training populations responding to their targeted product profiles. The end goal of these efforts is to make breeding more efficient through the routine use of genomic estimated breeding values to identify and select individuals in the breeding program to fast track for release, as well as use as parents for the next generation.

12.3 Modern Breeding Integration

12.3.1 Marker-Assisted Selection

Genomic-assisted breeding has been used widely in a number of plants and animals of importance in agriculture. It includes both marker-assisted breeding (MAB) and genomic selection-assisted breeding (GS) in making decisions regarding crossing, progeny selection, and yield trials. In polyploid crops, including sweetpotato, the presence of multiple genome copies introduces a high degree of complexity which in turn imposes numerous challenges to genome analysis and subsequent implementation in applied breeding programs (Kyriakidou et al. 2018; Ahmad et al. 2023).

Over the past few years, the polyploid community has achieved significant strides in bringing the genetics of polyploid species into the genomics era. The technological barrier of assessing these complex genomes got a breakthrough recently by use of high-throughput DNA sequencing technology. The delivery of massive amounts of DNA through such technology and their subsequent conversion into quantitative, dosage-dependent SNP-binary markers has enabled the analysis of these complex polyploid genomes. The massive amounts of information generated call for the need for powerful computational tools that are able to process raw DNA sequences to identify genetic markers. These markers are then used to construct linkage maps and infer haplotypes, identify the position of candidate genes through QTL mapping, and use the relationship between genotype and phenotype to make informed breeding decisions. This is what we refer to as genomic-assisted breeding in sweetpotato.

The polyploid community, which our group is a significant part of, has made significant advances in developing the tools needed for marker-assisted breeding in sweetpotato to be realized. Some of these tools include: SuperMASSA (Serang et al. 2012) and VCF2SM (Pereira et al. 2018), used for dosage-based variant calling in polyploids; MAPpoly (Mollinari et al. 2020) and polymapR (Bourke et al. 2018), used for integrated linkage map construction in a polyploid biparental population using dosage-based markers; QTLpoly (Da Silva et al. 2020), a software for performing QTL analysis in polyploid species, to mention but a few. QTLpoly combines the phenotypes of each individual in a mapped population with the genotype conditional probability distribution at each genomic position. Therefore, this software can be used to perform a variety of genetic analyses between phenotypes and genotypes including genomic selection and prediction. In selection of superior individuals for crossing and/or advancing in the breeding program, QTLpoly software can be extended beyond QTL analysis to predict genomic breeding values of clones (Da Silva et al. 2020).

We have conducted a number of studies using these new tools, some of which are published and summarized in Table 12.1. All these studies identified the genetic inheritance model of these traits as well as SNP markers that are linked to the traits. We are currently furthering this work by digging deeper into these identified QTL to identify low-cost usable markers for routine application in breeding programs.

12.3.2 Genomic Selection

In modern times, the primary objective of most breeding programs is to predict the genetic value of un-phenotyped individuals, enabling the targeted combination of desirable alleles to enhance the performance of future generations. The utilization of genomic selection (GS) in breeding programs has been demonstrated to effectively enhance genetic gains per unit of time, leading to the rapid identification of superior genotypes and acceleration of the breeding cycles (Heffner et al. 2010; Crossa et al. 2017).

However, the effective implementation of GS in crop breeding requires the utilization of prediction models that can improve the accuracy of predictions across diverse trait-environment combinations. One crucial aspect of plant breeding programs involves conducting multi-environment trials (MET), which aim to evaluate the performance of candidate genotypes under different environmental conditions (Jarquin et al. 2020). Furthermore, the genotype-by-environment (GE) interaction is an essential component of genetic variability, and a good understanding of this phenomenon can assist in identifying stable genotypes or genotypes with specific adaptations (Crossa et al. 2011; Hu et al. 2023).

Table 12.1 Breeding and genetic studies conducted using new breeding tools for polyploid sweetpotato

Genomic tools	Population	Crop	Trait	QTL identified	References
I. trifida genome, GBSpoly, MAPpoly, QTLpoly	Beauregard × Tanzania, 315 full sibs	Sweetpotato ($2n = 6x = 90$)	β-carotene and starch	2 QTL on LG 3 and 12	Gemenet et al. (2020a)
I. trifida genome, GBSpoly, MAPpoly, QTLpoly	Tanzania × Beauregard, 244 full sibs	Sweetpotato ($2n = 6x = 90$)	Southern RKN (*Meloidogyne Incognita*)	1 QTL on LG7	Oloka et al. (2021)
I. trifida genome, GBSpoly, MAPpoly, QTLpoly	Tanzania × Beauregard, 244 full sibs	Sweetpotato ($2n = 6x = 90$)	Guava RKN (*Meloidogyne enterolobii*)	1 QTL on LG4	Fraher (2022)

In breeding programs for clonally propagated species, such as sweetpotatoes, the impact of dominance on the effectiveness of GS implementation may be of critical importance, owing to the heterozygous nature of genotypes and the genetic value being a function of both additive and non-additive gene action (Gemenet et al. 2020b). Consequently, breeders are faced with the arduous task of increasing the additive value over time while simultaneously preserving the dominance value via the selection and recombination of parents (Werner et al. 2023). Batista et al. (2022) demonstrated in sweetpotato and sugarcane that if the trait has a high mean dominance degree and the population has a high frequency of heterozygous genotypes, the digenic dominance effects can significantly improve genomic prediction.

To identify superior parents for breeding programs or predict the potential of cross combinations, a comprehensive understanding of genetic architecture, accurate prediction of individual genetic values, and estimation of genetic parameters are indispensable. Yan et al. (2022) provided a comprehensive overview of the evolution of genetic and genomic tools for sweetpotato improvement, while Gemenet et al. (2020b) evaluated different strategies for genomic selection in this crop, highlighting their significant contributions to advancing genomic selection approaches in sweetpotato breeding. In this regard, a multiple-environment genomic best linear unbiased prediction (GBLUP) model that considers additive and dominance genetic effects can be helpful (Jarquín et al. 2014).

In hexaploid species, the additive matrix and digenic dominance matrices have been described by Batista et al. (2022). The relationship matrices can be calculated computationally using the R package AGHmatrix (Amadeu et al. 2016). The variance components, fixed and random effects in a statistical model are considered unknown quantities that need to be estimated and predicted. The restricted maximum likelihood (REML) method, as proposed by Patterson and Thompson (1971), can be used to estimate these variance components. The fixed effects represented as best linear unbiased estimates (BLUE), and random effects represented as best linear unbiased predictions (BLUP), can be estimated and predicted using mixed model equations (Henderson 1953). R packages such as 'sommer' (Covarrubias-Pazaran 2016) and ASReml (Butler et al. 2023) can be used to carry out these procedures.

The significance of the effects model and the quality of the model fit can be evaluated to understand the factors that impact the phenotypic traits and their genetic architectures. The significance of fixed effects can be assessed using a Wald test, while the significance of random effects can be evaluated using a likelihood ratio test (Luke 2017). Additionally, measures of model performance such as the Akaike information criterion (AIC) and the Bayesian information criterion (BIC) can be utilized to assess the goodness-of-fit of the model. Moreover, the success of genomic selection in breeding programs depends on the model's ability to predict the genetic value of un-phenotyped individuals with their genotypes recorded. The predictive ability of the model can be evaluated through cross-validation procedures using previously described populations and measures of predictive ability such as the Pearson correlation between phenotypes and genome-estimated breeding values.

12.3.3 Other Omics Technologies

'Omics' can be defined as different approaches that aim to measure biological molecules at particular levels through a large amount of data. Next-generation sequencing (NGS) techniques allow the generation of high throughput and fast nucleic acid data for different fields of biological research. The application of multiple 'omics' (multi-omics) techniques is essential for exploring the genetic roots of traits through genome composition (genomics), gene expression (transcriptomics), protein analyses (proteomics), and metabolites characterization (metabolomics). Several biological processes were elucidated in different plant species through multi-omics (Yang et al. 2021).

Being a highly heterozygous autohexaploid species with cross-incompatibility, genetic analysis, and molecular breeding of sweetpotato is a challenge. Compared to diploid and major polyploid crops, the 'omics' world of sweetpotato is lagging. The situation is mainly due to the complexity of the sweetpotato genome and the relatively small number of researchers working with the species (Yan et al. 2022). The utilization of multi-omics in sweetpotato genetic research could bypass the difficulties imposed by the species complexity and help advance sweetpotato molecular breeding.

12.3.3.1 Genomics

Genomics studies aim to uncover the full genetic content of an organism, employing DNA sequencing and bioinformatic methods to assemble the whole genome contents, identify genes, and determine their structures and functions. Analyses of a high-quality assembled genome can reveal genetic variations that affect desired phenotypes through the identification of sequence polymorphisms and chromosomal arrangements as well as understanding of how the regulation of gene expression affects phenotype expression (Yang et al. 2021). Several diploid species have their genomes fully sequenced and assembled; hence, usage and manipulation of their genomes are highly advanced. Unfortunately, polyploid species are relatively behind in those fields when compared to major diploid crops.

Autopolyploid genomes are usually hard to assemble due to the often high heterozygosity and high level of repetitive DNA in these genomes. For progressing with molecular breeding, a high-quality genome sequence is imperative. The first public draft genome for sweetpotato (Yang et al. 2017) generated 15 pseudochromosomes based on gene synteny with *Ipomoea nil* genome, and it is available at http://public-genomes-ngs.molgen.mpg.de/SweetPotato/. The assembly is very fragmented and contains a significant amount of redundancy and misassembly (Wu et al. 2018).

Due to the difficulties of assembling a polyploid genome, the generation of genomes from closely related diploid species is an alternative. There are five genomes from diploid *Ipomoea*

species available. *I. nil* (Hoshino et al. 2016), *I. purpurea* (Gupta et al. 2023) and *I. aquatica* (Hao et al. 2021). Although having good assembly quality overall, such genomes are phylogenetically distant from sweetpotato, which hinders their usage as reference genomes. On the other hand, *I. trifida* and *I. triloba* are closely related to *I. batatas*. In fact, *I. trifida* is considered to be a progenitor of sweetpotato (Roullier et al. 2013) and had its first tentative genome assembly performed by Hirakawa et al. (2015). The first chromosome-level reference genome for *I. trifida* together with *I. triloba* genome was published by Wu et al. (2018). The authors used a combination of genome sequencing, RNA sequencing (RNA-seq), molecular mapping, comparative genomics and predicted proteomes from different plant species to fully characterize the assemblies. The alignment of sweetpotato genomic sequences against both diploid genomes proved the genomes as good references for *I. batatas*, with more than 90% successfully aligned reads. Both genome sequences are available at https://sweetpotato.uga.edu/. A high-quality genome assembly was also constructed for a storage root forming *I. trifida* genotype (Y22). The analysis of the genome provided evidence of natural horizontal gene transfer from *Agrobacterium tumefaciens* to *I. trifida*. The *A. tumefaciens* sequence is also present in the sweetpotato genome, suggesting that sweetpotato might have inherited the sequence from diploid *I. trifida* (Lee et al. 2019).

Due to the quality of the *I. trifida* genome assembly and its similarity to sweetpotato genome, *I. trifida* genome sequence is being used as a reference genome for sweetpotato in a variety of research subjects such as development and validation of genetic tools for polyploids (Wadl et al. 2018; Mollinari et al. 2020; Yamakawa et al. 2021), linkage and QTL mapping (Da Silva et al. 2020; Oloka et al. 2021) and identification of candidate genes (Bararyenya et al. 2020; Gemenet et al. 2020b). As efforts to develop a high-quality assembly of the sweetpotato genome are ongoing, the availability of the *I. trifida* genome enabled advancements into sweetpotato genomic research.

12.3.3.2 Transcriptomics

The complete set of RNA transcripts, produced in cells and tissues, is called transcriptome and trancriptomics aims to characterize those transcript sets. Transcriptomics allows identification of putative genes, gene-targeted molecular markers, and genes differential expression profiles regarding different stimuli, time period, or developmental stage, being RNA-seq the most popular technique for transcriptome studies (Yang et al. 2021).

The first high-throughput sequencing of sweetpotato RNAs was performed by Schafleitner et al. (2010). A sweetpotato gene index was generated, and 24,657 putative unique genes were identified. The sequences were further used for short sequence repeats (SSRs) mining and 195 SSR markers were developed, which are used in sweetpotato breeding populations. Since then, transcriptome studies in *I. batatas* have increased noticeably, and this approach is the most used in sweetpotato molecular research. Transcriptomics was used to characterize gene expression in sweetpotato roots, primarily for the understanding of storage root formation (Wang et al. 2010; Firon et al. 2013; Ponniah et al. 2017), and defense response against root-knot nematode disease through transcriptional changes (Lee et al. 2019, 2021) as well as profile the gene expression in sweetpotato leaves under abiotic stress (Arisha et al. 2020a, b; Kitavi et al. 2023).

RNA-seq data from different tissues and organs of *I. trifida* and *I. triloba* were used for quality assessment and annotation of both species' genome assemblies (see above). In addition, the transcriptome of orange-fleshed cultivar 'Beauregard' was generated, and genes involved with storage root formation and β-carotene content were shown to present different gene expression regulation from their diploid species' counterparts (Wu et al. 2018). Using a combination of differential gene expression analysis, QTL mapping, and genome annotation data, Gemenet et al. (2020a) analyzed the negative association between β-carotene and starch accumulation, which is a result of starch

and carotenoid biosynthesis pathways competition for available carbon. Their results indicated that the physical proximity of *sucrose synthase* (*SuSY*) and *phytoene synthase* (*PSY*) genes affects the balance between pathways, while the *Orange* (*Or*) gene regulates *PSY* expression, acting as a molecular switch for carotenoid accumulation.

12.3.3.3 Proteomics

Proteomics profiles the total protein expression of an organism, analyzing their amino acid sequences, molecular structures, and functional activities. The ongoing improvements in protein extraction and purification protocols facilitated the advancements in the field. Protein data are generated by several methods such as high-performance liquid chromatography (HPLC), crystallization, X-ray diffraction of protein crystals, and yeast two-hybrid systems. In the last years, advanced high-throughput techniques using labeled amino acids with nuclear magnetic resonance (NMR) and mass spectrometry were developed (Yang et al. 2021). In sweetpotato, proteomics was used to compare protein expression profiles among genotypes of different flesh colors (Lee et al. 2012; Shekhar et al. 2016), allowing the detection of increased enzyme antioxidant activity and soluble sugar content in low-temperature storage roots (Cui et al. 2020), and identification of proteins involved in drought and heat simultaneous defense response (Tang et al. 2023).

Comparative proteomics was applied between pencil and storage roots to identify proteins that were up-regulated and/or uniquely expressed among both organ types. Pencil roots were overexpressing proteins related to cell wall, phenylpropanoid pathway, and antioxidant defense responses, indicating a higher rate of lignin biosynthesis and stress-related responses. On the other hand, proteins involved with development (maturity) and defense response against insects were up-regulated in storage roots. mRNA levels were in accordance with the protein expression results and lignin accumulation was only observed in the pencil roots. The authors argue that the carbon flow shift from the phenylpropanoid pathway to carbohydrate metabolism has major importance in storage root formation (Lee et al. 2015).

A broad profiling of the sweetpotato leaf and root organs proteome of the orange flesh cultivar 'Beauregard' was performed. In total, 74,255 peptides matching 4321 non-redundant proteins were identified and compared to available *Ipomoea* species predicted proteins, sweetpotato transcriptome, and genome sequences. More than 700 new coding regions were identified, and the analysis showed that approximately 2000 loci might be misannotated, showing the importance of using different methods to provide quality molecular information. Additionally, proteins unique to each organ were detected. Leaf proteins were mainly associated with primary metabolism and translation, which was associated with the growing activity and generation of metabolites and energy for the plant. Storage root was enriched for proteins involved with primary metabolism, intracellular transport, and protein localization, indicating the role of the organ as a nutrient sink (Al-Mohanna et al. 2019).

12.3.3.4 Metabolomics

Metabolites are small molecules, such as amino acids, lipids, and sugars that are intermediates or end products formed in metabolic processes. These molecules act as cellular structural constituents, building blocks for larger molecules, substrates for enzymatic reactions, and as signals for diverse signaling pathways (Baker and Rutter 2023). Metabolomics analyzes the metabolites that are involved with cellular processes of an organism, being the whole set of metabolites called metabolome. Each group type of metabolites has their own chemical/physical characteristics; therefore, analytical methods differ according to the wanted metabolic profile. The most used techniques are NMR and a combination of gas/liquid chromatography and mass spectrometry. For plants, which are the organisms that most produce metabolites, this field of research is especially important (Yang et al.

2021). Metabolomics studies in sweetpotato only started being reported in the last decade.

Metabolomics, together with proteomics, was used to characterize drought stress response in sweetpotato leaves. A drought-resistant cultivar showed the up-regulation of proteins involved in photosynthesis, reactive oxygen species (ROS) metabolism, and energy generation as well as the accumulation of carbohydrates, amino acids, flavonoids, and organic acids metabolites. Phenylpropanoid biosynthesis pathway-related proteins and metabolites were highly expressed and correlated, indicating a co-regulation among them (Zhou et al. 2022).

Carotenoid, flavonoid, anthocyanin, and phenolic acid metabolites were profiled in white, orange, and purple-fleshed sweetpotato roots. Orange-fleshed sweetpotatoes had the highest level of carotenoids, while white-fleshed genotypes had the lowest level. Flavonoid concentration was higher in purple-fleshed sweetpotato, followed by orange- and white-fleshed varieties. Anthocyanins were virtually only present in the purple-fleshed genotypes, in which six phenolic acid levels are tenfold higher than in white- and orange-fleshed samples. In addition, orange- and purple-fleshed sweetpotatoes had higher concentrations of sugars and sugar alcohols (Park et al. 2016; Wang et al. 2018).

A joint analysis of transcriptome and metabolome data from different flesh-colored sweetpotatoes showed that the regulatory network of anthocyanin production in sweetpotato roots involves not only specific anthocyanins biosynthetic genes as the process is also highly regulated by the flux allocation and modification of metabolites. In addition, the *flavonol synthase* (*FLS*) gene was shown to be crucial for the regulation of anthocyanin biosynthesis. Purple-fleshed sweetpotatoes have lower expression of *FLS*, when compared to white- and orange-fleshed roots, which leads to the different pigmentation among the cultivars (Xiao et al. 2023).

The understanding of metabolite content changes during sweetpotato processing is important as well since the roots are usually cooked before consumption. Heating raised the polyphenol content in the cooked roots and antioxidant activity was higher when compared to raw samples. Antioxidant activity was highly correlated to chlorogenic acids content, which was also enhanced in the cooked sweetpotatoes (Franková et al. 2022).

In addition, cooking promotes the saccharification of sweetpotato roots, a process in which sugars are produced through hydrolysis or acidolysis of starch or cellulose. Maltose is the primary contributor to sweetpotato sweetness after cooking since its content increases sharply, while other sugar metabolites' content shows no significant change. Using transcriptome and metabolome information, Lee et al. (2021) identified *starch synthase* (*SS*), *granular starch synthase* (*GBSS*), and *branching enzyme* (*GBE*) genes as regulators of the transformation of starch to maltose probably by regulating starch structure. The metabolites identified were enriched by starch and sucrose metabolic pathways. Nevertheless, some of the metabolites had no annotation although being correlated with annotated genes.

12.4 Final Remarks

Due to the polyploid nature and genetic complexity of sweetpotato, the need for genomic-assisted breeding is of significance in accelerating its breeding process. We have developed and applied new genetic tools to identify QTL as well as understand the genetic architecture of a number of important traits in sweetpotato. These new tools, which include the fully sequenced and annotated diploid lines *I. trifida* and *I. triloba*, which serve as a reference genome for hexaploid sweetpotato, are steps in the right direction to realize faster genetic gains in sweetpotato and deliver it in farmers' fields in the form of improved resilient and nutritious varieties.

There are a few technical challenges that we will need to consider and overcome to fully exploit the benefits of these tools, which include the complex family structure of polyploid breeding populations. Typically, this family structure has been in the form of multiple partially

inter-related half-sib families coming from a polycross block of about 20 parents. In these populations, only the female parent is known, at best because there are instances where seeds from polycross nurseries are bulked. This family structure has been used by the sweetpotato community for decades in part due to the significantly high levels of cross-incompatibility in sweetpotato, the difficulty in identifying superior individuals in early breeding generations, and the challenge in generating enough seed through targeted crosses.

Genomics-assisted breeding will not remove all these existing hurdles, but it will considerably reduce the need for making multiple 'blind' crosses from many parents with unknown genetic backgrounds. It will also eliminate the need to evaluate thousands of early generation clones in multiple environments before realizing their potential as either parents or potential new varieties. When used hand in hand with advanced statistical methodologies and analytics, GS in sweetpotato would allow for evaluation of individuals predicted to have high breeding values for selecting traits of importance in multiple environments at earlier generations. This will greatly accelerate breeding prospects and have a long-term effect of lowering the cost for releasing new varieties to replace old ones in farmers' fields, thereby increasing genetic gain for traits of interest.

References

Agrios G (2004) Plant pathology, 5th edn. Elsevier, Amsterdam

Ahmad N, Fatima S, Mehmood MA, Zaman QU, Atif RM, Zhou W, Rahman M, Gill RA (2023) Targeted genome editing in polyploids: lessons from *Brassica*. Front Plant Sci 14:1152468

Akinsanya A, Afolami S, Kulakow P, Parkes E, Coyne D (2020) Popular biofortified cassava cultivars are heavily impacted by plant parasitic nematodes, especially *Meloidogyne* Spp. Plants 9(6):802. https://doi.org/10.3390/plants9060802

Al-Mohanna T, Ahsan N, Bokros NT, Dimlioglu G, Reddy KR, Shankle M, Popescu GV, Popescu SC (2019) Proteomics and proteogenomics analysis of sweetpotato (*Ipomoea batatas*) leaf and root. J Proteome Res 18(7):2719–2734. https://doi.org/10.1021/acs.jproteome.8b00943

Amadeu RR, Cellon C, Olmstead JW, Garcia AAF, Resende MFR, Muñoz PR (2016) AGHmatrix: R package to construct relationship matrices for autotetraploid and diploid species: a blueberry example. Plant Genome 9(3):9. https://doi.org/10.3835/plantgenome2016.01.0009

Amankwaah VA (2019) Phenotyping and genetic studies of storage root chemistry traits in sweetpotato. North Carolina State University, Raleigh

Anyanga MO, Muyinza H, Talwana H, Hall DR, Farman DI, Ssemakula GN, Mwanga ROM, Stevenson PC (2013) Resistance to the weevils *C. puncticollis* and *C. brunneus* conferred by sweetpotato root surface compounds. J Agric Food Chem 61(34):8141–8147. https://doi.org/10.1021/jf4024992

Anyanga MO, Yada B, Yencho GC, Ssemakula GN, Alajo A, Farman DI, Mwanga ROM, Stevenson PC (2017) Segregation of hydroxycinnamic acid esters mediating sweetpotato weevil resistance in storage roots of sweetpotato. Front Plant Sci 8(6):1–8. https://doi.org/10.3389/fpls.2017.01011

Arisha MH, Aboelnasr H, Ahmad MQ, Liu Y, Tang W, Gao R, Yan H, Kou M, Wang X, Zhang Y, Li Q (2020a) Transcriptome sequencing and whole genome expression profiling of hexaploid sweetpotato under salt stress. BMC Genom 21(1):197. https://doi.org/10.1186/s12864-020-6524-1

Arisha MH, Ahmad MQ, Tang W, Liu Y, Yan H, Kou M, Wang X, Zhang Y, Li Q (2020b) RNA-sequencing analysis revealed genes associated drought stress responses of different durations in hexaploid sweet potato. Sci Rep 10(1):12573. https://doi.org/10.1038/s41598-020-69232-3

Aritua V, Bua B, Barg E, Vetten HJ, Adipala E, Gibson RW (2007) Incidence of five viruses infecting sweetpotatoes in Uganda; the first evidence of Sweet potato caulimo-like virus in Africa. Plant Pathol 56(2):324–331. https://doi.org/10.1111/j.1365-3059.2006.01560.x

Baker SA, Rutter J (2023) Metabolites as signalling molecules. Nat Rev Mol Cell Biol 24(5):355–374. https://doi.org/10.1038/s41580-022-00572-w

Bararyenya A, Olukolu BA, Tukamuhabwa P, Grüneberg WJ, Ekaya W, Low J, Ochwo-Ssemakula M, Odong TL, Talwana H, Badji A, Kyalo M, Nasser Y, Gemenet D, Kitavi M, Mwanga ROM (2020) Genome-wide association study identified candidate genes controlling continuous storage root formation and bulking in hexaploid sweetpotato. BMC Plant Biol 20(1):3. https://doi.org/10.1186/s12870-019-2217-9

Batista LG, Mello VH, Souza AP, Margarido GRA (2022) Genomic prediction with allele dosage information in highly polyploid species. Theor Appl Genet 135(2):723–739. https://doi.org/10.1007/s00122-021-03994-w

Bernardo R, Yu J (2007) Prospects for genomewide selection for quantitative traits in maize. Crop Sci 47(3):1082–1090. https://doi.org/10.2135/cropsci2006.11.0690

Bourke PM, Voorrips RE, Visser RGF, Maliepaard C (2018) Tools for genetic studies in experimental populations of polyploids. Front Plant Sci 9:513. https://doi.org/10.3389/fpls.2018.00513

Butler DG, Cullis BR, Gilmour AR, Gogel BJ, Thompson R (2023) ASReml-R reference manual version 4.2. VSN International Ltd., Hemel Hempstead

Chitwood DJ (2003) Nematicides. In: Encyclopedia of agrochemicals. Wiley, Hoboken

Clark CA, Davis JA, Abad JA, Cuellar WJ, Fuentes S, Kreuze JF, Gibson RW, Mukasa SB, Tugume AK, Tairo FD, Valkonen JPT (2012) Sweetpotato viruses: 15 years of progress on understanding and managing complex diseases. Plant Dis 96(2):168–185. https://doi.org/10.1094/PDIS-07-11-0550

Covarrubias-Pazaran G (2016) Genome-assisted prediction of quantitative traits using the R Package sommer. PLoS ONE 11(6):e0156744. https://doi.org/10.1371/journal.pone.0156744

Coyne DL, Kagoda F, Wambugu E, Ragama P (2006) Response of cassava to nematicide application and plant-parasitic nematode infection in East Africa, with emphasis on root knot nematodes. Int J Pest Manag 52(3):215–223. https://doi.org/10.1080/09670870600722959

Crossa J, Pérez-Rodríguez P, Cuevas J, Montesinos-López O, Jarquín D, Campos G, Burgueño J, González-Camacho JM, Pérez-Elizalde S, Beyene Y, Dreisigacker S, Singh R, Zhang X, Gowda M, Roorkiwal M, Rutkoski J, Varshney RK (2017) Genomic selection in plant breeding: methods, models, and perspectives. Trends Plant Sci 22(11):961–975. https://doi.org/10.1016/j.tplants.2017.08.011

Crossa J, Pérez P, de los Campos G, Mahuku G, Dreisigacker S, Magorokosho C (2011) Genomic selection and prediction in plant breeding. J Crop Improv 25(3):239–261. https://doi.org/10.1080/15427528.2011.558767

Cui P, Li Y, Cui C, Huo Y, Lu G, Yang H (2020) Proteomic and metabolic profile analysis of low-temperature storage responses in *Ipomoea batata* Lam. tuberous roots. BMC Plant Biol 20(1):435. https://doi.org/10.1186/s12870-020-02642-7

Da Silva PG, Gemenet DC, Mollinari M, Olukolu BA, Wood JC, Diaz F, Mosquera V, Gruneberg WJ, Khan A, Buell CR, Yencho GC, Zeng Z-B (2020) Multiple QTL mapping in autopolyploids: a random-effect model approach with application in a hexaploid sweetpotato full-sib population. Genetics 215(3):579–595. https://doi.org/10.1534/genetics.120.303080

Downham MCA, Smit NEJM, Laboke PO, Hall DR, Odongo B (2001) Reduction of pre-harvest infestations of African sweetpotato weevils *C. brunneus* and *C. puncticollis* (Coleoptera: Apionidae) using a pheromone mating-disruption technique. Crop Prot 20:353–353. https://doi.org/10.1016/S0261-2194(01)00024-2

Duhnen A, Gras A, Teyssèdre S, Romestant M, Claustres B, Daydé J, Mangin B (2017) Genomic selection for yield and seed protein content in soybean: a study of breeding program data and assessment of prediction accuracy. Crop Sci 57(3):1325–1337. https://doi.org/10.2135/cropsci2016.06.0496

Firon N, LaBonte D, Villordon A, Kfir Y, Solis J, Lapis E, Perlman T, Doron-Faigenboim A, Hetzroni A, Althan L, Nadir L (2013) Transcriptional profiling of sweetpotato (*I. batatas*) roots indicates down-regulation of lignin biosynthesis and up-regulation of starch biosynthesis at an early stage of storage root formation. BMC Genom 14(1):460. https://doi.org/10.1186/1471-2164-14-460

Fraher SP (2022) Advancing molecular tools for the accelerated release of root-knot nematode resistant sweetpotato varieties. NC State University, Raleigh

Franková H, Musilová J, Árvay J, Šnirc M, Jančo I, Lidiková J, Vollmannová A (2022) Changes in antioxidant properties and phenolics in sweet potatoes (*I. batatas L.*) due to heat treatments. Molecules 27(6):1884. https://doi.org/10.3390/molecules27061884

Gemenet DC, da Silva Pereira G, De Boeck B, Wood JC, Mollinari M, Olukolu BA, Diaz F, Mosquera V, Ssali RT, David M, Kitavi MN, Burgos G, Felde TZ, Ghislain M, Carey E, Swanckaert J, Coin LJM, Fei Z, Hamilton JP, Yada B, Yencho GC, Zeng Z-B, Mwanga ROM, Khan A, Gruneberg WJ, Buell CR (2020a) Quantitative trait loci and differential gene expression analyses reveal the genetic basis for negatively associated β-carotene and starch content in hexaploid sweetpotato [*I. batatas* (L.) Lam.]. Theor Appl Genet 133(1):23–36. https://doi.org/10.1007/s00122-019-03437-7

Gemenet DC, Lindqvist-Kreuze H, De Boeck B, da Silva PG, Mollinari M, Zeng Z-B, Craig Yencho G, Campos H (2020b) Sequencing depth and genotype quality: accuracy and breeding operation considerations for genomic selection applications in autopolyploid crops. Theor Appl Genet 133(12):3345–3363. https://doi.org/10.1007/s00122-020-03673-2

Gibson RW, Mpembe I, Alical T, Carey EE, Mwanga ROM, Seal SE, Vetten HJ (1998) Symptoms, aetiology and serological analysis of sweet potato virus disease in Uganda. Plant Pathol 47(1):95–102. https://doi.org/10.1046/j.1365-3059.1998.00196.x

Grüneberg WJ, Eyzaguirre R, Díaz F, Boeck B, Espinoza J, Swanckaert J, Dapaah H, Andrade MI, Makunde GS, Agili S, Ndingo-Chipungu FP, Attaluri S, Kapinga R, Nguyen T, Kaiyung X, Tjintokohadi K, Ssali RT, Carey T, Low JW, Mwanga ROM (2019) Procedures for the evaluation of sweetpotato trials. International Potato Center

Gupta S, Harkess A, Soble A, Van Etten M, Leebens-Mack J, Baucom RS (2023) Interchromosomal linkage disequilibrium and linked fitness cost loci associated with selection for herbicide resistance. New Phytol 238(3):1263–1277. https://doi.org/10.1111/nph.18782

Hao Y, Bao W, Li G, Gagoshidze Z, Shu H, Yang Z, Cheng S, Zhu G, Wang Z (2021) The chromosome-based genome provides insights into the evolution in water spinach. Sci Hortic 289:110501. https://doi.org/10.1016/j.scienta.2021.110501

Heffner EL, Sorrells ME, Jannink JL (2009) Genomic selection for crop improvement. Crop Sci 49(1):1–12. https://doi.org/10.2135/cropsci2008.08.0512

Heffner EL, Lorenz AJ, Jannink J-L, Sorrells ME (2010) Plant breeding with genomic selection: gain per unit time and cost. Crop Sci 50(5):1681–1690. https://doi.org/10.2135/cropsci2009.11.0662

Henderson CR (1953) Estimation of variance and covariance components. Biometrics 9(2):226–252. https://doi.org/10.2307/3001853

Hirakawa H, Okada Y, Tabuchi H, Shirasawa K, Watanabe A, Tsuruoka H, Minami C, Nakayama S, Sasamoto S, Kohara M, Kishida Y, Fujishiro T, Kato M, Nanri K, Komaki A, Yoshinaga M, Takahata Y, Tanaka M, Tabata S, Isobe SN (2015) Survey of genome sequences in a wild sweet potato, *Ipomoea trifida* (H. B. K.) G. Don. DNA Res 22(2):171–179. https://doi.org/10.1093/dnares/dsv002

Hoshino A, Jayakumar V, Nitasaka E, Toyoda A, Noguchi H, Itoh T, Shin IT, Minakuchi Y, Koda Y, Nagano AJ, Yasugi M, Honjo MN, Kudoh H, Seki M, Kamiya A, Shiraki T, Carninci P, Asamizu E, Nishide H, Tanaka S, Park K-I, Morita Y, Yokoyama K, Uchiyama I, Tanaka Y, Tabata S, Shinozaki K, Hayashizaki Y, Kohara Y, Suzuki Y, Sugano S, Fujiyama A, Iida S, Sakakibara Y (2016) Genome sequence and analysis of the Japanese morning glory *Ipomoea nil*. Nat Commun 7(1):13295. https://doi.org/10.1038/ncomms13295

Hu X, Carver BF, El-Kassaby YA, Zhu L, Chen C (2023) Weighted kernels improve multi-environment genomic prediction. Heredity 130(2):82–91. https://doi.org/10.1038/s41437-022-00582-6

Jarquín D, Kocak K, Posadas L, Hyma K, Jedlicka J, Graef G, Lorenz A (2014) Genotyping by sequencing for genomic prediction in a soybean breeding population. BMC Genom 15(1):740. https://doi.org/10.1186/1471-2164-15-740

Jarquin D, Howard R, Crossa J, Beyene Y, Gowda M, Martini JWR, Covarrubias Pazaran G, Burgueño J, Pacheco A, Grondona M, Wimmer V, Prasanna BM (2020) Genomic prediction enhanced sparse testing for multi-environment trials. G3 GenesGenomesGenetics 10(8):2725–2739. https://doi.org/10.1534/g3.120.401349

Jones JT, Haegeman A, Danchin EGJ, Gaur HS, Helder J, Jones MGK, Kikuchi T, Manzanilla-López R, Palomares-Rius JE, Wesemael WML, Perry RN (2013) Top 10 plant-parasitic nematodes in molecular plant pathology. Mol Plant Pathol 14(9):946–961. https://doi.org/10.1111/mpp.12057

Karuri HW, Olago D, Neilson R, Mararo E, Villinger J (2017) A survey of root knot nematodes and resistance to *Meloidogyne incognita* in sweet potato varieties from Kenyan fields. Crop Prot 92:114–121. https://doi.org/10.1016/j.cropro.2016.10.020

Karyeija RF, Kreuze JF, Gibson RW, Valkonen JPT (2000) Synergistic interactions of a potyvirus and a phloem-limited crinivirus in sweet potato plants. Virology 269(1):26–36. https://doi.org/10.1006/viro.1999.0169

Kitavi M, Gemenet DC, Wood JC, Hamilton JP, Wu S, Fei Z, Khan A, Buell CR (2023) Identification of genes associated with abiotic stress tolerance in sweetpotato using weighted gene co-expression network analysis. Plant Direct 7(10):e532. https://doi.org/10.1002/pld3.532

Kyriakidou M, Tai HH, Anglin NL, Ellis D, Strömvik MV (2018) Current strategies of polyploid plant genome sequence assembly. Front Plant Sci 9:1660. https://doi.org/10.3389/fpls.2018.01660

Lee JJ, Park KW, Kwak Y-S, Ahn JY, Jung YH, Lee B-H, Jeong JC, Lee H-S, Kwak S-S (2012) Comparative proteomic study between tuberous roots of light orange- and purple-fleshed sweetpotato cultivars. Plant Sci 193–194:120–129. https://doi.org/10.1016/j.plantsci.2012.06.003

Lee JJ, Kim Y-H, Kwak Y-S, An JY, Kim PJ, Lee BH, Kumar V, Park KW, Chang ES, Jeong JC, Lee H-S, Kwak S-S (2015) A comparative study of proteomic differences between pencil and storage roots of sweetpotato (*I. batatas* (L.) Lam.). Plant Physiol Biochem 87:92–101. https://doi.org/10.1016/j.plaphy.2014.12.010

Lee IH, Shim D, Jeong JC, Sung YW, Nam KJ, Yang J-W, Ha J, Lee JJ, Kim Y-H (2019) Transcriptome analysis of root-knot nematode (*Meloidogyne incognita*)-resistant and susceptible sweetpotato cultivars. Planta 249(2):431–444. https://doi.org/10.1007/s00425-018-3001-z

Lee I-H, Kim HS, Nam KJ, Lee K-L, Yang J-W, Kwak S-S, Lee JJ, Shim D, Kim Y-H (2021) The defense response involved in sweetpotato resistance to root-knot nematode *Meloidogyne incognita*: comparison of root transcriptomes of resistant and susceptible sweetpotato cultivars with respect to induced and constitutive defense responses. Front Plant Sci 12:671677. https://doi.org/10.3389/fpls.2021.671677

Luke SG (2017) Evaluating significance in linear mixed-effects models in R. Behav Res Methods 49(4):1494–1502. https://doi.org/10.3758/s13428-016-0809-y

Mcharo M, LaBonte D, Mwanga ROM, Kriegner A (2005) Associating molecular markers with virus resistance to classify sweetpotato genotypes. J Am Soc Hortic Sci 130(3):355–359. https://doi.org/10.21273/JASHS.130.3.355

Miano DW, LaBonte DR, Clark CA (2008) Identification of molecular markers associated with sweet potato resistance to sweet potato virus disease in Kenya. Euphytica 160(1):15–24. https://doi.org/10.1007/s10681-007-9495-2

Mollinari M, Olukolu BA, Pereira GDS, Khan A, Gemenet D, Yencho GC, Zeng Z-B (2020) Unraveling the hexaploid sweetpotato inheritance using ultra-dense multilocus mapping. G3 GenesGenomesGenetics 10(1):281–292. https://doi.org/10.1534/g3.119.400620

Montesinos-López A, Montesinos-López OA, Gianola D, Crossa J, Hernández-Suárez CM (2018) Multi-environment genomic prediction of plant traits using deep learners with dense architecture. G3 GenesGenomesGenetics 8(12):3813–3828. https://doi.org/10.1534/g3.118.200740

Morales N, Ogbonna AC, Ellerbrock BJ, Bauchet GJ, Tantikanjana T, Tecle IY, Powell AF, Lyon D, Menda N, Simoes CC, Saha S, Hosmani P, Flores M, Panitz N, Preble RS, Agbona A, Rabbi I, Kulakow P, Peteti P, Kawuki R, Esuma W, Kanaabi M, Chelangat DM, Uba E, Olojede A, Onyeka J, Shah T, Karanja M, Egesi C, Tufan H, Paterne A, Asfaw A, Jannink J-L, Wolfe M, Birkett CL, Waring DJ, Hershberger JM, Gore MA, Robbins KR, Rife T, Courtney C, Poland J, Arnaud E, Laporte M-A, Kulembeka H, Salum K, Mrema E, Brown A, Bayo S, Uwimana B, Akech V, Yencho C, de Boeck B, Campos H, Swennen R, Edwards JD, Mueller LA (2022) Breedbase: a digital ecosystem for modern plant breeding. G3 Genes Genomes Genet 12(7):jkac078. https://doi.org/10.1093/g3journal/jkac078

Mwanga ROM, Mayanja S, Swanckaert J, Nakitto M, Felde T, Grüneberg W, Mudege N, Moyo M, Banda L, Tinyiro SE, Kisakye S, Bamwirire D, Anena B, Bouniol A, Magala DB, Yada B, Carey E, Andrade M, Johanningmeier SD, Forsythe L, Fliedel G, Muzhingi T (2021a) Development of a food product profile for boiled and steamed sweetpotato in Uganda for effective breeding. Int J Food Sci Technol 56(3):1385–1398. https://doi.org/10.1111/ijfs.14792

Mwanga ROM, Swanckaert J, da Silva PG, Andrade MI, Makunde G, Grüneberg WJ, Kreuze J, David M, De Boeck B, Carey E, Ssali RT, Utoblo O, Gemenet D, Anyanga MO, Yada B, Chelangat DM, Oloka B, Mtunda K, Chiona M, Koussao S, Laurie S, Campos H, Yencho GC, Low JW (2021b) Breeding progress for vitamin A, iron and zinc biofortification, drought tolerance, and sweetpotato virus disease resistance in sweetpotato. Front Sustain Food Syst 5:6674. https://doi.org/10.3389/fsufs.2021.616674

Mwanga ROM, Odongo B, P'Obwoya CO, Gibson RW, Smit NEJM, Carey EE (2001) Release of five sweetpotato cultivars in Uganda. HortScience 36(2):385–386. https://doi.org/10.21273/HORTSCI.36.2.385

Mwanga ROM, Kriegner A, Cervantes-Flores JC, Zhang DP, Moyer JW, Yencho GC (2002) Resistance to sweetpotato chlorotic stunt virus and sweetpotato feathery mottle virus is mediated by two separate recessive genes in sweetpotato. J Am Soc Hortic Sci 127(5):798–806. https://doi.org/10.21273/JASHS.127.5.798

Nottingham SF, Son K-C, Wilson DD, Severson RF, Kays SJ (1989) Feeding and oviposition preferences of sweet potato weevil, *Cylas formicarius elegantulus* (Summers), on storage roots of sweet potato cultivars with differing surface chemistries. J Chem Ecol 15(3):895–903. https://doi.org/10.1007/BF01015185

Oloka BM (2018) Genetic linkage map construction and QTL analysis of important pest and agronomic traits in two bi-parental sweetpotato SNP mapping populations. North Carolina State University, Raleigh

Oloka BM, da Silva PG, Amankwaah VA, Mollinari M, Pecota KV, Yada B, Olukolu BA, Zeng Z-B, Craig Yencho G (2021) Discovery of a major QTL for root-knot nematode (*Meloidogyne incognita*) resistance in cultivated sweetpotato (*I. batatas*). Theor Appl Genet 134(7):1945–1955. https://doi.org/10.1007/s00122-021-03797-z

Ozimati A, Kawuki R, Esuma W, Kayondo SI, Pariyo A, Wolfe M, Jannink J-L (2019) Genetic variation and trait correlations in an east African cassava breeding population for genomic selection. Crop Sci 59(2):460–473. https://doi.org/10.2135/cropsci2018.01.0060

Park S-Y, Lee SY, Yang JW, Lee J-S, Oh S-D, Oh S, Lee SM, Lim M-H, Park SK, Jang J-S, Cho HS, Yeo Y (2016) Comparative analysis of phytochemicals and polar metabolites from colored sweet potato (*I. batatas* L.) tubers. Food Sci Biotechnol 25(1):283–291. https://doi.org/10.1007/s10068-016-0041-7

Patterson HD, Thompson R (1971) Recovery of inter-block information when block sizes are unequal. Biometrika 58(3):545–554. https://doi.org/10.1093/biomet/58.3.545

Pereira GS, Garcia AAF, Margarido GRA (2018) A fully automated pipeline for quantitative genotype calling from next generation sequencing data in autopolyploids. BMC Bioinform 19(1):398. https://doi.org/10.1186/s12859-018-2433-6

Pipitpukdee S, Attavanich W, Bejranonda S (2020) Impact of climate change on land use, yield and production of cassava in Thailand. Agriculture 10(9):402. https://doi.org/10.3390/agriculture10090402

Ponniah SK, Thimmapuram J, Bhide K, Kalavacharla V, Manoharan M (2017) Comparative analysis of the root transcriptomes of cultivated sweetpotato (*I. batatas* [L.] Lam) and its wild ancestor (*Ipomoea trifida* [Kunth] G. Don). BMC Plant Biol 17(1):9. https://doi.org/10.1186/s12870-016-0950-x

Resende MDV, Resende MFR, Sansaloni CP, Petroli CD, Missiaggia AA, Aguiar AM, Abad JM, Takahashi EK, Rosado AM, Faria DA, Pappas GJ, Kilian A, Grattapaglia D (2012) Genomic selection for growth and wood quality in Eucalyptus: capturing the missing heritability and accelerating breeding for complex traits in forest trees. New Phytol 194(1):116–128. https://doi.org/10.1111/j.1469-8137.2011.04038.x

Roullier C, Duputié A, Wennekes P, Benoit L, Fernández Bringas VM, Rossel G, Tay D, McKey D, Lebot V (2013) Disentangling the origins of cultivated sweet potato (*I. batatas* (L.) Lam.). PLoS ONE 8(5):e62707. https://doi.org/10.1371/journal.pone.0062707

Sasser JN (1980) Root-knot nematodes: a globe. Plant Dis 64(1):36. https://doi.org/10.1094/PD-64-36

Schafleitner R, Tincopa LR, Palomino O, Rossel G, Robles RF, Alagon R, Rivera C, Quispe C, Rojas L, Pacheco JA, Solis J, Cerna D, Young Kim J, Hou J, Simon R (2010) A sweetpotato gene index established by de novo assembly of pyrosequencing and Sanger sequences and mining for gene-based microsatellite markers. BMC Genom 11(1):604. https://doi.org/10.1186/1471-2164-11-604

Serang O, Mollinari M, Garcia AAF (2012) Efficient exact maximum a posteriori computation for bayesian SNP genotyping in Polyploids. PLoS ONE 7(2):e30906. https://doi.org/10.1371/journal.pone.0030906

Shekhar S, Mishra D, Gayali S, Buragohain AK, Chakraborty S, Chakraborty N (2016) Comparison of proteomic and metabolomic profiles of two contrasting ecotypes of sweetpotato (*Ipomoea batata* L.). J Proteom 143:306–317. https://doi.org/10.1016/j.jprot.2016.03.028

Stevenson PC, Muyinza H, Hall DR, Porter E a, Farman DI, Talwana H, Mwanga ROM (2009) Chemical basis for resistance in sweetpotato *I. batatas* to the sweetpotato weevil *C. puncticollis*. Pure Appl Chem 81(1):141–151. https://doi.org/10.1351/PAC-CON-08-02-10

Tairo F, Mukasa SB, Jones R a. C, Kullaya A, Rubaihayo PR, Valkonen JPT (2005) Unravelling the genetic diversity of the three main viruses involved in sweet potato virus disease (SPVD), and its practical implications. Mol Plant Pathol 6(2):199–211. https://doi.org/10.1111/j.1364-3703.2005.00267.x

Tang W, Arisha MH, Zhang Z, Yan H, Kou M, Song W, Li C, Gao R, Ma M, Wang X, Zhang Y, Li Z, Li Q (2023) Comparative transcriptomic and proteomic analysis reveals common molecular factors responsive to heat and drought stresses in sweetpotaoto (*I. batatas*). Front Plant Sci 13:1081948. https://doi.org/10.3389/fpls.2022.1081948

Wadl PA, Olukolu BA, Branham SE, Jarret RL, Yencho GC, Jackson DM (2018) Genetic diversity and population structure of the USDA sweetpotato (*I. batatas*) germplasm collections using GBSpoly. Front Plant Sci 9:1166. https://doi.org/10.3389/fpls.2018.01166

Wang Z, Fang B, Chen J, Zhang X, Luo Z, Huang L, Chen X, Li Y (2010) De novo assembly and characterization of root transcriptome using Illumina paired-end sequencing and development of cSSR markers in sweetpotato (*I. batatas*). BMC Genom 11(1):726. https://doi.org/10.1186/1471-2164-11-726

Wang A, Li R, Ren L, Gao X, Zhang Y, Ma Z, Ma D, Luo Y (2018) A comparative metabolomics study of flavonoids in sweet potato with different flesh colors (*I. batatas* (L.) Lam). Food Chem 260:124–134. https://doi.org/10.1016/j.foodchem.2018.03.125

Werner CR, Gaynor RC, Sargent DJ, Lillo A, Gorjanc G, Hickey JM (2023) Genomic selection strategies for clonally propagated crops. Theor Appl Genet 136(4):74. https://doi.org/10.1007/s00122-023-04300-6

Wu S, Lau KH, Cao Q, Hamilton JP, Sun H, Zhou C, Eserman L, Gemenet DC, Olukolu BA, Wang H, Crisovan E, Godden GT, Jiao C, Wang X, Kitavi M, Manrique-Carpintero N, Vaillancourt B, Wiegert-Rininger K, Yang X, Bao K, Schaff J, Kreuze J, Gruneberg W, Khan A, Ghislain M, Ma D, Jiang J, Mwanga ROM, Leebens-Mack J, Coin LJM, Yencho GC, Buell CR, Fei Z (2018) Genome sequences of two diploid wild relatives of cultivated sweetpotato reveal targets for genetic improvement. Nat Commun 9(1):4580. https://doi.org/10.1038/s41467-018-06983-8

Xiao J, Xu X, Li M, Wu X, Guo H (2023) Regulatory network characterization of anthocyanin metabolites in purple sweetpotato via joint transcriptomics and metabolomics. Front Plant Sci 14:1030236. https://doi.org/10.3389/fpls.2023.1030236

Yada B, Alajo A, Ssemakula GN, Brown-Guedira G, Otema MA, Stevenson PC, Mwanga ROM, Craig Yencho G (2017a) Identification of simple sequence repeat markers for sweetpotato weevil resistance. Euphytica 213(6):129. https://doi.org/10.1007/s10681-017-1917-1

Yada B, Alajo A, Ssemakula GN, Mwanga ROM, Brown-Guedira G, Yencho GC (2017b) Selection of simple sequence repeat markers associated with inheritance of sweetpotato virus disease resistance in sweetpotato. Crop Sci 57(3):1421. https://doi.org/10.2135/cropsci2016.08.0695

Yada B (2014) Genetic analysis of agronomic traits and resistance to sweetpotato weevil and sweet potato virus disease in a bi-parental sweetpotato population. Raleigh, North Carolina: North Carolina State University, pp 1. online resource (xii, 203 pages): illustrations. Available at: http://www.lib.ncsu.edu/resolver/1840.16/9557

Yada B, Brown-Guedira G, Alajo A, Ssemakula GN, Mwanga ROM, Yencho GC (2015) Simple sequence repeat marker analysis of genetic diversity among progeny of a biparental mapping population of sweetpotato. HortScience 50(8):1143–1147. https://doi.org/10.21273/HORTSCI.50.8.1143

Yamakawa H, Haque E, Tanaka M, Takagi H, Asano K, Shimosaka E, Akai K, Okamoto S, Katayama K, Tamiya S (2021) Polyploid QTL-seq towards rapid development of tightly linked DNA markers for potato and sweetpotato breeding through whole-genome resequencing. Plant Biotechnol J 19(10):2040–2051. https://doi.org/10.1111/pbi.13633

Yan M, Nie H, Wang Y, Wang X, Jarret R, Zhao J, Wang H, Yang J (2022) Exploring and exploiting genetics and genomics for sweetpotato improvement: Status and perspectives. Plant Commun 3(5):100332. https://doi.org/10.1016/j.xplc.2022.100332

Yang J, Moeinzadeh M-H, Kuhl H, Helmuth J, Xiao P, Haas S, Liu G, Zheng J, Sun Z, Fan W, Deng G, Wang H, Hu F, Zhao S, Fernie AR, Boerno S, Timmermann B, Zhang P, Vingron M (2017) Haplotype-resolved sweet potato genome traces back its hexaploidization history. Nat Plants 3(9):696–703. https://doi.org/10.1038/s41477-017-0002-z

Yang Y, Saand MA, Huang L, Abdelaal WB, Zhang J, Wu Y, Li J, Sirohi MH, Wang F (2021) Applications of multi-omics technologies for crop improvement. Front Plant Sci 12:563953. https://doi.org/10.3389/fpls.2021.563953

Zhou Z, Tang J, Cao Q, Li Z, Ma D (2022) Differential response of physiology and metabolic response to drought stress in different sweetpotato cultivars. PLoS ONE 17(3):e0264847. https://doi.org/10.1371/journal.pone.0264847

Index

© The Editor(s) (if applicable) and The Author(s) 2025
G. C. Yencho et al. (eds.), *The Sweetpotato Genome,* Compendium of Plant Genomes,
https://doi.org/10.1007/978-3-031-65003-1